Pollution Is Colonialism

Duke University Press *Durham and London* 2021

Pollution Is Colonialism

MAX LIBOIRON

Printed in the United States of America on acid-free paper ∞
Project Editor: Lisl Hampton
Designed by Courtney Leigh Richardson
Typeset in Garamond Premier Pro and Scala Sans Pro by
Copperline Book Services

Library of Congress Cataloging-in-Publication Data
Names: Liboiron, Max, [date] author.
Title: Pollution is colonialism / Max Liboiron.
Description: Durham : Duke University Press, 2021. | Includes bibliographical references and index.
Identifiers: LCCN 2020037439 (print)
LCCN 2020037440 (ebook)
ISBN 9781478013228 (hardcover)
ISBN 9781478014133 (paperback)
ISBN 9781478021445 (ebook)
Subjects: LCSH: Environmentalism—Political aspects. | Imperialism—Environmental aspects. | Pollution—Political aspects. | Pollution—Social aspects. | Research—Environmental aspects. | Research—Political aspects. | Traditional ecological knowledge. | Environmental justice.
Classification: LCC JA75.8.L54 2021 (print) | LCC JA75.8 (ebook) | DDC 304.2/8—dc23
LC record available at https://lccn.loc.gov/2020037439
LC ebook record available at https://lccn.loc.gov/2020037440

Cover art: From the series *Dinner Plates (Northern Fulmar)*, 2015.
Photo by Max Liboiron. Courtesy the artist.

Contents

Acknowledgements

The territory in which this text was written is the ancestral homeland of the Beothuk. The island of Newfoundland is the ancestral homelands of the Mi'kmaq and Beothuk. I would also like to recognize the Inuit of Nunatsiavut and NunatuKavut and the Innu of Nitassinan, and their ancestors, as the original people of Labrador. We strive for respectful relationships with all the peoples of this province as we search for collective healing and true reconciliation and honour this beautiful land together.[1]

Taanishi. Max Liboiron dishinihkaashoon. Lac la Biche, Treaty siz, d'ooshchiin. Métis naasyoon, niiya ni: nutr faamii Woodman, Turner, pi Umperville awa. Ni papaa (kii ootinikaatew) Jerome Liboiron, pi ni mamaa (kii ootinikaatew) Lori Thompson. Ma paaraan et Richard Chavolla (Kumeyaay). I'm from Lac la Biche, Treaty 6 territory in northern Alberta, Canada. The parents who raised me are Jerome Liboiron and Lori Thompson. I am connecting with Métis family through a lineage of Woodman, Turner, and Umperville that leads back to Red River. Rick Chavolla of Kumeyaay Nation is my godfather. These are my guiding relations.[2]

1 This Land acknowledgement was created collectively with leaders of most of the province's Indigenous governing bodies. These are not my words: they are words chosen for guests of this land. They are not mine to change.

2 Dear Reader: thank you for being here. Introductions are important because they show where my knowledge comes from, to whom I am accountable, and how I was built. Some of these things are not for promiscuous, public consumption and some of them are. To young Indigenous thinkers in particular: one of my struggles has been how to introduce myself properly without also telling stories that aren't mine to share or that feature personal or

In his first year, PhD student Edward Allen came into my office, sat down in a small wooden chair that was certainly not built for him, and asked if his name had to be on his dissertation. He argued that because his dissertation would be a product of many people's knowledge, putting his name on the front page would be a misrepresentation of authorship. I am fortunate to keep such company. His point is a good one: no intellectual work is authored alone. Many people built this book. Many are acknowledged here and throughout the text in footnotes so readers can see whose shoulders I stand on. I see these footnotes enacting an ethic of gratitude, acknowledgement, and reciprocity for their work. They make it harder to imagine these words are just mine, an uninterrupted monologue. They are not stashed at the end, but physically interrupt the text to support it and show my relationships. Here, footnotes build a world full of thinkers whom I respect. By putting footnotes on the page, I aim to account for how citations are "*screening techniques*: how certain bodies take up spaces by screening out the existence of others," as well as "reproductive technolog[ies], a way of reproducing the world around certain bodies."[3] Citing the knowledges of Black, Indigenous, POC, women, LGBTQAI+, two-spirit, and young thinkers is one small part of an anticolonial methodology that refuses to reproduce the myth that knowledge, and particularly science, is the domain of pale, male, and stale gatekeepers.

Building a book reminds me of what Alex Wilson (Cree) calls "coming in," or coming to better understand our "relationship to and place and value in [our] own family, community, culture, history, and present-day world."[4] Wilson is talking about coming in as a two-spirit process of place-based relationality, in

---

familial trauma or scandal. I have tried to model the ways mentors have taught me to introduce myself that point up to structures of relation or oppression rather than pointing down to effects on family. Charismatic as the practice may be, I will never open a vein to bleed for my audience to make the case that colonial violence exists in our everyday lives. I have watched so many of you at conferences talk about your traumas and your pain, often to make the case that our intellectual labour has stakes, has roots, has validity, has teeth. Personally and professionally, I don't think academic spaces have earned that blood. I watch the (mostly white) academic audiences at these talks become rapt and feel the pleasure and the depth of blood-and-trauma talk, but I also think that these arguments are only heard in a way that allows many to continue to believe that Indigenous people are inherently traumatized, always already bleeding. Charisma, after all, is about resonance with existing values and ideas. It is your choice how you introduce yourself. One of my guideposts for introduction and the place of trauma is Tuck, "Suspending Damage."

3 S. Ahmed, "Making Feminist Points." For more on the politics of citation, see Mott and Cockayne, "Citation Matters"; and Tuck, Yang, and Gaztambide-Fernández, "Citation Practices."

4 A. Wilson, "N'tacinowin Inna Nah," 196.

juxtaposition to various LGBTQAI+ ways of coming out as a declaration of self. Writing a book queerly, two-spiritly, is (I think and feel) an act of coming in, circling back to belonging, sharing-in, and the accountabilities that come with that, much of which is done in the footnotes.

TL;DR: My goal is to do science differently. Part of that is happening in the footnotes.

I suspect that the first[5] person I should acknowledge is the one I text in the middle of the day because I've come to an irreconcilable tension in the book's argument, and who gives up her time to talk me through it, not as charity or even friendship (though that, too), but as part of a lesson in good relations and familyhood. Emily Simmonds, I hope you see how your teaching by example is reflected in every aspect of the book. Thank you, and Constance, for the place you've given me in your—our—family. Maarsi.

Likewise, Rick Chavolla has been teaching me about relations, ethics, and bold statements for years. He was teaching me back when he was just a baby Elder. Thank you, Rick, for supporting me so I could choose not to drop out of my PhD and for adopting me into your family as a goddaughter. You've taught me about prayer and how important core muscles are for running away from the police. Same lesson. Part of me always lives on your couch. Thank you, Rick Chavolla and Anna Ortega-Chavolla. I hope you understand how this book is the way it is because of you, both in its detail and in the broad strokes. This is what your love can do.

This book is about relations, and no one has taught me more about good relations through everyday lessons than Grandmother and Kookum. Almost every word of this book has been written within ten feet of Grandmother, which is a blessing all on its own.

I cannot thank and acknowledge Michelle Murphy enough. Michelle, your intellectual, emotional, familial, and pedagogical labours have influenced my thinking and the way I relate to others as a scholar. You are part of my ability to flourish in academia. From the first time we met on a panel at 4S and you cau-

5 There are ways to do acknowledgements that refuse to order people. Andrea Ballestero's *A Future History of Water* is a beautiful example, where acknowledgements are in a kaleidoscope of relations. It is cool, smart, and refuses structures that prioritize, rank, categorize. In my scientific work with CLEAR (whom you'll meet in a moment), we talk about the hierarchy of ordering a lot, and we choose to stay with the tensions of ordering for many reasons. Here, I order my thanks because the way I have been taught obligation does indeed prioritize some over others. For more on ordering ethically, or at least equitably and with humility, see M. Liboiron et al., "Equity in Author Order."

tioned me about fetishizing molecules (I totally was), to emergency Skype calls during my first year as a professor when I wanted to either burn academia down or quit for good, to reminding me to be more kind and less hard-edged, to your presence at the birth of this book, you are and have been my academic auntie.[6] Thank you, so much. Love is an insufficient term to characterize your mentorship and friendship.

Reaching all the way back to the people who taught me early lessons about relations and who gave me (and continue to give me) support to go to that mythical place called university and do that stuff called academia (though we didn't know the term at the time), thank you, Lori, Jerome, Joel, Curtis, and Melissa, as well as Mila. I have been adopted into several families in the last decade, but you are my first and forever family. You are the ground I stand on. Without you, I could not take the risks I can because I know I can only fall so far.

Lessons in relations are done in place. Gratitude to Lac la Biche, Edmonton, New York City, and Newfoundland and Labrador for sharing lessons and correcting my ignorance and hubris regularly. Maarsi.

In different but overlapping ways, Alisa Craig[7] and Nicole Power made it possible for me to stay on, work in, and learn from the island of Newfoundland. I can't imagine what it would be like to be an academic here without you. I would likely not be here, and certainly I would not be as smart, funny, content, or successful as I am (or think I am) without you. Thank you, both. Thank you especially to Nicole for blending our families and supporting the logistics of family that fills out my life to something fuller than I could have imagined before. Likewise, thank you to Neil Bose and the VPR team for enabling a way of working and doing good in the province and university that I could never have done alone, and certainly do not want to do alone. When someone has your back the way you have mine, things become possible that were unimaginable before. Thank you.

Readers, did you know there is this wonderful type of event where people who are invested in you and your work come together, on couches and over food, for a couple of days to give feedback on your book? I didn't, until Joe Masco told me about it. I think it has a real name, but I've called it a book doula party. It means peer review is based in love and generosity—one of the greatest academic gifts I have ever received. To the book doulas who took time out of their busy lives to hold the head of this book and guide it into the world—Michelle

6 For more on academic aunties, see E. Lee, "I'm Concerned for Your Academic Career."

7 Alisa Craig is the star department chair mentioned in Liboiron, "Care and Solidarity Are Conditions for Interventionist Research."

Murphy, Emily Simmonds, Rick Chavolla, Joe Masco, and Nick Shapiro—thank you for taking time and care and, most of all, responsibility for ensuring this book grew up on a good path.

Thank you to the people who make my world, a writing world, possible: MUNFA, my faculty union; the custodial staff and facilities management personnel in the Arts Building, Science Building, and Bruneau Centre at Memorial who were the first people to officially welcome me to my new role in Bruneau and who watched Grandmother grow up (and still check on his health); Arn Keeling and Shannon Fraser for looking after the family; to Heather O'Brien, Matt Milner, Kelley Bromley-Brits, Bradley Cooper, Carrie Dyke, Ruby Bishop, and Sharon Roseman for frontline work on red tape and incorrectly calculated Excel sheets; and a special thanks to Pam Murphy, without whom the list of Civic Laboratory for Environmental Action Research (CLEAR) members below would be much, much shorter.

The most interesting, exciting, frustrating, essential, confusing, fulfilling, difficult, and beautiful aspects of my research would not be possible without CLEAR members, past and present. First, thank you to present and past (undergraduate!) lab managers who directly improve my quality of life and the quality of lived experiences in the lab: France Liboiron (no blood relation, all lab relation), Natasha Healey, and, most recently and magnificently, Kaitlyn Hawkins. The other voices of CLEAR, in no particular order, are Emily Wells, Hillary Bradshaw, Tristen Morris, Melissa Novachefski, Emily Simmonds, Natalya Dawe, Coco Coyle, Mikayla Downey, Erin Burt, Juddyannet Murichi, Natalie Richárd, Jackie Saturno, Charles Mather, Nicole Power, Marissa Van Harmelen, Elise Earles, Jess Melvin, Taylor Stocks, Alexandra Hayward, Justine Ammendolia, Bojan Fürst, John Atkinson, Kate Winsor, Lucas Harris, Sam Welscott, Ignace Schoot, Nadia Duman, Alex Zahara, Edward Allen, Nic Kuzmochka, Taylor Hess, Noah Hutton, Shramana Sarkar, Kelechi Anyaeto, Luke Lucy-Broomfield, Megan Dicker, Charlotte Muise, Jillian Chidley, Carley Mills, Lauren Watwood, Ayo Oladimeji Awalaye, Molly Rivers, Tiaasha Naskar, Celestine Muli, Domenica Lombeida, Michael Bros, Alexander Flynn, Tammy Sheppard, Zhe Shi, Forough Emam, Christina Crespo, Arif Abu, and Melissa Paglia, as well as Louis Charron. Thank you for your collective thinking, writing, experimenting, counting, eating, smelling, sieving, working remotely, and risk-taking that has defined CLEAR for the first five years of its life. The lab could not have existed when and how it did without the generosity of Yolanda Wiersma lending her storage closet as a lab when we first began, and without Stephanie Avery-Gomm, Michelle Valliant, Carley Schacter, Katherine Robins, Ian Jones, and all those dovekies to get things started.

A special thanks to Emily Simmonds, Natalya Dawe, and Reena Shadaan for your work on compiling and polishing literature reviews that have either heavily influenced or appeared in this text.

For conversations about toxicity, pollution, plastics, research ethics, place-based methodologies, models of collegiality, and other meaty things that have nourished my thinking and learning, thank you: Liz Pijogge, Tina Ngata, Alex Wilson, Catharyn Andersen, Chris Andersen (different Andersens), Kelly Ann Butler, Kim TallBear, Jesse Jacobs, Jamie Snook, Carla Pamak, Ashlee Cunsolo, Sarah Martin, Dvera Saxton, Shannon Dosemagen, Nick Shapiro, Beza Merid, Katie Pine, Luke Stark, Erica Robles-Anderson, Rochelle Gutierrez, Katherine Crocker, Marisa Duarte, Shannon Fraser, Chelsea Rochman, Rebecca Altman, Athena LaTocha, Manuel Tironi, Nerea Calvillo, Robin Nagle, Samantha MacBride, Josh Lepawsky, Jen Henderson, Sarah Wylie, Arn Keeling, Alex Zahara, Neil Bose, Carissa Brown, Dean Bavington, Kim Fortun, Scott Knowles, Phoebe Sengers, Joe Masco, Desi Rodriquez-Lonebear, Gary Downey, Teun Zuiderent-Jerak, Stephanie Russo Carroll, Paul McCarney, Rodd Laing, Shannon Cram, Jack Tchen, Ken Paul, Daniel Cohen, David Wachsmuth, Shelly Ronen, Marisa Solomon, Vincent Lai, Aaron Bornstein, Cecil Scheib, Song Chong, Lawrence Mesich, Ellen Steinhauer, Kyle Powys Whyte, Scott Neilson, Carlin Wing, Blithe Riley, Caitlin Wells, Nicole Burisch, Heather O'Brien, Sergio Sismondo, Alex Fink, Catherine Kenny, Carmella Gray-Cosgrove, Donny Persaud, Christy Spackman, Natasha Myers, Jenny Molloy, Lisa Diedrich, Audra Wolfe, Anna Cummins, Marcus Eriksen, Carolynn Box, and, of course (and again), Michelle Murphy and Emily Simmonds. Citations show how and where I stand on your shoulders, but many of you don't produce knowledge in a way that Google Scholar can identify (Google's loss). I've tried to leave your tracks in the text in ways you can recognize. Undoubtedly, I've missed some people. That sucks and I'm sorry.

The first blush of this book started in my dissertation, and its tracks are left in the first chapter. I thank my New York University PhD committee and advisors Robin Nagle, Lisa Gitelman, Erica Robles-Anderson, and especially Brett Gary. A rocky start and an attempt to flee turned into a home run with your guidance.

Many of the ideas in this book were worked out in conferences, panels, Q&As, and hallway chats within my various research communities, including the Geography Department at Memorial University, the Society for Social Studies of Science (4S), Native American and Indigenous Studies Association (NAISA) conferences, the Indigenous STS network, the Endocrine Disruptors Action Group (EDAction), the Gathering for Open Science Hardware (GOSH), the Chemical

Heritage Foundation (now the Science History Institute), the WaSTE group, the Toxic Legacies gathering, and Superstorm Research Lab.

The archival research, land and water work, literature reviews, and other data gathering for this book, as well as editing and writing support, were funded by the Social Sciences and Humanities Research Council of Canada (SSHRC) (#435-2017-0567 and #430-2015-00413); the Marine Environmental Observation, Prediction and Response Network (MEOPAR) (who took a risk and became my first scientific funder!); the Northern Contaminants Program out of Indian and Northern Affairs Canada; ArcticNet; the Seed, Bridge, and Multidisciplinary Fund from Memorial University; the Office of the Vice President (Research) at Memorial; and start-up funds from the Department of Sociology and the Faculty of Humanities and Social Sciences (and, as I realize now, the Office of the Vice President [Research]!). Chapter 2 benefitted from an Allington Residency Fellowship at the Chemical Heritage Foundation and an Anna K. and Mary E. Cunningham Research Residency at New York State Library in Albany.

Thank you, all, for building this book.

Introduction

In 1956, Lloyd Stouffer, the editor of the US magazine *Modern Packaging*, addressed attendees at the Society of the Plastics Industry meeting in New York City: "The future of plastics is in the trash can. . . . It [is] time for the plastics industry to stop thinking about 'reuse' packages and concentrate on single use. For the package that is used once and thrown away, like a tin can or a paper carton, represents not a one-shot market for a few thousand units, but an everyday recurring market measured by the billions of units."[1] Stouffer was speaking at a time when reuse, making do, and thrift were key practices reinforced by two US wars. Consumer markets were saturating. Disposability was one tactic within a suite of efforts to move goods *through*, rather than merely *into*, consumer households.[2] Today, packaging is the single largest category of plastic production, ac-

1 Hello, Reader! Thank you for being here. These footnotes are a place of nuance and politics, where the protocols of gratitude and recognition play out (sometimes also called citation), where warnings and care work are carried out (including calling certain readers aside for a chat or a joke), and where I contextualize, expand, and emplace work. The footnotes support the text above, representing the shoulders on which I stand and the relations I want to build. They are part of doing good relations within a text, through a text. Since a main goal of *Pollution Is Colonialism* is to show how methodology is a way of being in the world and that ways of being are tied up in obligation, these footnotes are one way to enact that argument. Thank you to Duke University Press for these footnotes.

For this first footnote of the introduction, we have a simple citation: Stouffer, "Plastics Packaging," 1–3. Don't worry. They'll get better.

2 Packard, *Waste Makers*; Strasser, *Waste and Want*; M. Liboiron, "Modern Waste as Strategy."

counting for nearly 40 percent of plastic production in Europe[3] and 33 percent in Canada.[4] The next largest categories are building and construction, at just over 20 percent, and automotive at 8 percent.[5] Stouffer's desire looks like prophecy. (Spoiler: It isn't. It's colonialism, but more on that in a moment.)

Before Stouffer's call for disposability and before German and US military powers invested significant finances and research infrastructure into perfecting plastics as a wartime material in the 1940s, plastic was described as an environmental good.[6] Mimicking first ivory and then other animal-based materials such as shellac and tortoiseshell, plastic was an artisan substance that showcased technological ingenuity and skill while providing "the elephant, the tortoise, and the coral insect a respite in their native haunts; it will no longer be necessary to ransack the earth in pursuit of substances which are constantly growing scarcer."[7] The idea of disposability and mass production for plastics is relatively new, developing half a century after plastics were invented. Most plastic production graphs start their timelines after 1950, ignoring the nineteenth- and early

3 PlasticsEurope, "Plastics," 12. These numbers include thermoplastics and polyurethanes as well as thermosets, adhesives, coatings, and sealants, but they do not include PET, PA, PP, and polyacryl-fibers. Note that PET and PP are some of the most common plastics found in marine environments.

4 Deloitte and Cheminfo Services, "Economic Study of the Canadian Plastic Industry, Markets, and Waste," 6.

5 PlasticsEurope, "Plastics," 12.

6 While historian Jeffrey Meikle (unmarked, see below) provides much archival evidence on how plastics were written about as a replacement for animal products, it is not clear whether there were "actual" material shortages or not, nor is it clear whether plastics played a role in alleviating that shortage (or not). Regardless, this idea was still core to the early reputation of plastics. Meikle, *American Plastic*. For an alternative, see Friedel, *Pioneer Plastic*, 60–64. Thank you, Rebecca Altman (settler), for not only sharing this insight but also consistently prioritizing the work of others in such a way that you reach out as a co-thinker when people (like me) reproduce an academic truism that needs some empirical work. Thank you for your collegiality, for the way you celebrate other people's work with genuine enthusiasm and care, and for your careful chemical storytelling. Folks, see Altman, "Time-Bombing the Future"; Altman, "American Petro-Topia"; and Altman, "Letter to America."

Pioneer and *plastic* appear together quite a bit in both historical and present-day texts. While I will talk about plastic production's assumption of terra nullius, I won't dwell on its relationships to pioneering frontierism, except to say that the use of *pioneer* to mean innovation simultaneously normalizes frontierism and the forms of erasure, dispossession, and death frontierism requires to make its terra nullius.

7 Meikle, *American Plastic*, 12.

twentieth-century histories of plastics since these materials did not exist as the mass-produced substances we know today.[8] Plastics have been otherwise.

In 1960, only four years after Stouffer's address, a British ornithology journal published an account of the "confounding" discovery of a rubber band in a puffin's stomach.[9] It would be among the first of hundreds of published reports of wildlife ingesting plastics, including the ones I publish as an environmental scientist. How did plastics become such a ubiquitous pollutant? There are questions that should precede that question: What do you mean by pollutant? How did pollutants come to make sense in the first place? It turns out that the concept of environmental pollution as we understand it today is also new.

Only twenty years before Stouffer launched the future of plastics into the trash can, the now-dominant and even standard understanding of modern environmental pollution was articulated on the Ohio River. Two engineers in the brand-new field of sanitation engineering named Earle B. Phelps and H. W. Streeter (both unmarked)[10] created a scientific and mathematical model of the

8 See, e.g., PlasticsEurope, "Plastics," 12.

9 Bennett, "Rubber Bands in a Puffin's Stomach," 222.

10 It is common to introduce Indigenous authors with their nation/affiliation, while settler and white scholars almost always remain unmarked, like "Lloyd Stouffer." This unmarking is one act among many that re-centres settlers and whiteness as an unexceptional norm, while deviations have to be marked and named. Simone de Beauvoir (French) called this positionality both "positive and the neutral, as is indicated by the common use of *man* to designate human beings in general." Not cool. This led me to a methodological dilemma. Do I mark everyone? No one? I thought about just leaving it, because this is difficult and even uncomfortable to figure out, but since this is a methods text I figured I should shit or get off the pot. Feminist standpoint theory and even truth and reconciliation processes maintain that social location and the different collectives we are part of matter to relations, obligations, ethics, and knowledge. Settlers have a different place in reconciliation than Indigenous people, than Black people who were stolen from their Land. As la paperson (diasporic settler of colour) writes, "'Settler' is not an identity; it is the idealized juridical space of exceptional rights granted to normative settler citizens and the idealized exceptionalism by which the settler state exerts its sovereignty. The 'settler' is a site of exception from which whiteness emerges. . . . [T]he anthropocentric normal is written in its image." This assumed positive and neutral "normal" right is enacted in the lack of introduction of settlers as settlers, as if settler presence on Land, especially Indigenous Land, is the stable and unremarkable norm. What allows settlers to consistently and unthinkingly not introduce their relations to Land and colonial systems is settlerism. See paperson, *A Third University Is Possible*, 10; and Beauvoir, *Second Sex*.

In light of this complex terrain, my imperfect methodological decision has been to identify all authors the way they identify themselves (thank you to everyone who does this!) the first time they appear in a chapter. If an author does not introduce themselves

conditions and rates under which water (or at least that bit of the Ohio River) could purify itself of organic pollutants.[11] After running tests that accounted for different temperatures, velocities of water, concentrations of pollutants, and other variables, they wrote that self-purification is a "measurable phenomenon governed by definite laws and proceeding according to certain fundamental physical and biochemical reactions. Because of the fundamental character of these reactions and laws, it is fairly evident that the principles underlying the phenomenon [of self-purification] as a whole are applicable to virtually all polluted streams."[12]

The Streeter-Phelps equation, as it came to be known, not only became a hallmark of water pollution science and regulation but also contained within it their theory of pollution: that a moment existed when water could not purify itself and that moment could be measured, predicted, and properly called pollution. Self-purification became known as *assimilative capacity*,[13] a term of art

---

or their land relations, I mark them as "unmarked." I do this rather than marking settlers as settlers because of the advice of Kim TallBear (Sisseton-Wahpeton Oyate), who encourages people to look at structures of the settler state rather than focusing on naming individual settlers, which reenacts the logics of eugenicist and racist impulses to properly and finally categorize people properly. TallBear, Callison, and Harp. "Ep. 198."

I take up this method so we, as users of texts, can understand where authors are speaking from, what ground they stand on, whom their obligations are to, what forms of sovereignty are being leveraged, what structures of privilege the settler state affords, and how we are related so that our obligations to one another as speaker and listener, writer and audience, can be *specific enough to enact obligations to one another*, a key goal of this text. How has colonialism affected us differently? Introducing yourself is part of ethics and obligation, not punishment. Following Marisa Duarte's (Yaqui) example in *Network Sovereignty*, I simply introduce people in this way by using parentheses after the first time their name is mentioned. Duarte, *Network Sovereignty*.

11 Organic pollutants can also be industrial pollutants. Organic in this case does not mean naturally occurring—even arsenic, radon, and methylmercury, while "naturally occurring" compounds, do not occur in the tonnages and associated scales of toxicity without industrial infrastructure.

12 Streeter and Phelps, *Study of the Pollution and Natural Purification of the Ohio River*, 59.

13 Cognate terms that describe thresholds of harm used in different countries and contexts include *carrying capacity*, *critical load*, *allowable threshold*, and *maximum permissible dose*. Versions of the term in specific scientific disciplines include *reference dose* (RfD), *no observable adverse effect level* (NOAEL), *lowest observable adverse effect level* (LOAEL), *lethal dose 50 percent* (LD50), *median effective concentration* (EC50), *maximum acceptable concentration* (MAC), and *derived minimal effect level* (DMEL) (which is a truly tricky measure for a level of exposure for which the risk levels of a nonthreshold carcinogen become

in both environmental science and policy making that refers to "the amount of waste material that may be discharged into a receiving water without causing deleterious ecological effects."[14] State-based environmental regulations in most of the world since the 1930s are premised on the logic of assimilative capacity, in which a body—water, human, or otherwise—can handle a certain amount of contaminant before scientifically detectable harm occurs. I call this the threshold theory of pollution.

Plastics do not assimilate in the way that Streeter and Phelps's organic pollution assimilated in the Ohio River. As I pull little pieces of burned plastic out of a dovekie[15] gizzard in my marine science lab, the Civic Laboratory for Environmental Action Research (CLEAR), the threshold theory of pollution and the future of plastics as waste look like bad relations. I don't mean the individualized bad relations of littering (which does not produce much waste compared to other flows of plastic into the ocean, especially here in Newfoundland and Labrador, a land of fishing gear and untreated sewage) or the bad relations of capitalism where growth and profit are put before environmental costs (though those are certainly horrible relations). I mean the bad relations of a scientific theory that allows some amount of pollution to occur and its accompanying entitlement to Land to assimilate that pollution.[16] I mean colonialism.

The structures that allow plastics' global distribution and full integration into ecosystems and everyday human lives are based on colonial land relations, the assumed access by settler and colonial projects to Indigenous lands for settler and colonial goals. At the same time, the ways in which plastics pollute unevenly, do not follow threshold theories of harm, and act as both hosts for life and sources of harm have made plastics an ideal case to change dominant colonial concepts of pollution by teaching us about relations and obligations that

"tolerable," thus creating a social threshold where there are no toxicological thresholds). Each has different specifics, but the same theory lies behind them. More on this in chapter 1.

14 Novotny and Krenkel, "Waste Assimilative Capacity Model," 604.

15 A dovekie is also called a bully bird, little auk, or *Alle alle*, depending on who's talking. They look like tiny puffins without the fancy beak, and you can see them flying over the water in lines. Some people in Newfoundland and Labrador eat them, but the bones are tiny, thin, and hard to pick out.

16 This argument also appears in CLEAR and EDAction, "Pollution Is Colonialism," and is expanded beautifully in Shadaan and Murphy, "Endocrine-Disrupting Chemicals as Industrial and Settler Colonial Structures." Also see Ngata and Liboiron, "Māori Plastic Pollution Expertise."

tend to be obfuscated from view by environmental rhetoric and industrial infrastructures. In CLEAR, we place land relations at the centre[17] of our knowledge production as we monitor plastic pollution in the province of Newfoundland and Labrador.

As members of a marine science lab, we are dedicated to doing science differently by foregrounding *anti*colonial land relations. This requires critique but mostly it requires action.[18] We've stopped using toxic chemicals to process samples, which means there is a whole realm of analysis we can't do. We also use judgmental sampling rather than random sampling in our study design to foreground food sovereignty when we look at plastics in food webs. CLEAR does good with pollution, in practice, in place. But CLEAR is not unique: land relations always already play a central role in all sciences, anticolonial and otherwise.

I find that many people understand colonialism as a monolithic structure with roots exclusively in historical bad action, rather than as a set of contemporary and evolving land relations that can be maintained by good intentions and even good deeds. The call for more recycling, for example, still assumes access to Indigenous Land for recycling centres and their pollution. Other people have nuanced understandings of colonialism and seek ways to deal with colonial structures in their everyday lives and research, often in spaces like the academy that reproduce colonialism in uneven ways. This book is for both groups, and others besides. Overall, this is a methodological text that begins with colonial land relations, so that we can recognize them in familiar and comfortable places (like reading, like counting), and then considers anticolonial methods that centre and change colonial land relations in thought and action.

I make three main arguments in this book. First, pollution is not a manifestation or side effect of colonialism but is rather an enactment of ongoing colonial relations to Land.[19] That is, pollution is best understood as the violence of colo-

17 Perhaps you've noticed Canadian spellings in the text even though Duke University Press is based in the United States. This is a constant, possibly annoying, reminder that these words come from a place. Spelling is method.

18 Hale, "Activist Research v. Cultural Critique."

19 Throughout this book, you'll notice that sometimes *Land* is capitalized, and sometimes it isn't. I follow the lead of Styres and Zinga (Indigenous and settler, respectively), who "capitalize Land when we are referring to it as a proper name indicating a primary relationship rather than when used in a more general sense. For us, land (the more general term) refers to landscapes as a fixed geographical and physical space that includes earth, rocks, and waterways; whereas, 'Land' (the proper name) extends beyond a material fixed space. Land is a spiritually infused place grounded in interconnected and interdependent relationships, cultural positioning, and is highly contextualized" (300–301). Likewise, when I capitalize

nial land relations rather than environmental damage, which is a symptom of violence. These colonial relations are reproduced through even well-intentioned environmental science and activism. Second, there are ways to do pollution action, particularly environmental science, through different Land relations, and they're already happening without waiting for the decolonial horizon to appear. These methods are specific, place-based, and attend to obligations. Third, I show how methodologies—whether scientific, writerly, readerly, or otherwise—are always already part of Land relations and thus are a key site in which to enact good relations (sometimes called ethics). This last point should carry to a variety of contexts that do not focus on either pollution or the natural sciences.

I use the case of plastics, increasingly understood as an environmental scourge and something to be annihilated, to refute and refuse the colonial in a good way. That is, I try to keep plastics and pollution from being conflated too readily, instead decoupling them so existing and potential relations can come to light that exceed the popular position of "plastics are bad!"—even though plastics are often bad. To start, let's dig into colonialism (spoiler: it is not synonymous with "bad" in general, though it is certainly bad).

Colonialism

Stouffer, Streeter, and Phelps all assumed access to Indigenous Land when they made their proclamations. Stouffer's declaration about the future of plastics as disposables assumed that household waste would be picked up and taken

---

Land I am referring to the unique entity that is the combined living spirit of plants, animals, air, water, humans, histories, and events recognized by many Indigenous communities. When *land* is not capitalized, I am referring to the concept from a colonial worldview whereby landscapes are common, universal, and everywhere, even with great variation. For the same reason, I also capitalize *Nature* and *Resource* and, occasionally, *Science*. Rather than use a small *N* or *R* or *S* that might indicate that these words are common or universal, the capitalization signals that they are proper nouns that are highly specific to one place, time, and culture. That is, Nature is not universal or common, but unique to a specific worldview that came about at a particular time for specific reasons. Calling out proper nouns so they are also proper names is part of a tradition where using someone/thing's name is to bring it out of the shadows and engage it—in power, in challenge, in recognition, in kinship. That's why I don't mind looking like an academic elitist or naive literary wannabe when I capitalize. There's more on compromise in chapter 3. Styres and Zinga, "Community-First Land-Centred Theoretical Framework," 300–301. For other politics of capitalization in feminist sciences, see Subramaniam and Willey, "Introduction"; and Harding, *Science and Social Inequality*.

to landfills or recycling plants that allowed plastic disposables to go "away."[20] Without this infrastructural access to Indigenous Land, there is no disposability.[21] He assumed that Land would provide a sink, a place to store waste, so that profits could be generated through flows of waste-as-consumer-goods. This assumption is made easier when the Land has already been cleared of Indigenous peoples via genocide, moves to reserves, and ongoing disappearances such as those catalogued under MMIWG[22] statistics.

Streeter and Phelps likewise assumed access to Indigenous Land, though they were not capitalists dedicated to growth and profit. On the contrary, Phelps was a bold environmental conservationist. Unlike his contemporaries, he believed polluted rivers could and should be saved from, rather than abandoned to, industrial pollution by using science to keep the pollution be-

20 There is some excellent work on the concept of waste and its "away," including Davies, "Slow Violence and Toxic Geographies" and de Coverly et al., "Hidden Mountain."

21 I first made this argument in *Teen Vogue*: M. Liboiron, "How Plastic Is a Function of Colonialism." This is not the first and will not be the last time I cite myself. There are good reasons to self-cite in certain ways. First, in the words of fish philosopher Zoe Todd (Métis): "It is cheeky to cite oneself and to return to the same stories repeatedly in Euro-western academe. We are taught, as students and apprentices, that this is verboten (a well-meaning mentor even cautioned not to waste my good stories on the wrong journal, which is generally good advice for Euro-Western scholars). . . . However, Leroy Little Bear (Blackfoot) ['Big Thinking'] reminds us that 'in Native ways, we always retell our stories, we repeat them. That's how they sink in and become embodied in students and in the people.' It is through returning to the fish stories shared with me by interlocutors in Paulatuuq, and by reengaging the fish stories my family and friends share with me in amiskwaciwâskahikan, that I am brought back into my reciprocal relationships to people, moments, and responsibilities both in my research and in my engagement as a citizen of my home territory. By returning to the same moments time and time again, I unravel new facets of the relationships these stories contain and enliven." Todd, "Refracting the State," 61; Little Bear, "Big Thinking." Maarsi, Zoe Todd, for the work you do reorienting academics to good relations and manners. I admire the pedagogy your work uses to shore up unlearning and learning in the academy.

Second, I still happen to agree with myself on this point. That doesn't always happen. As I learn, I change my mind. Citing myself in specific ways marks where theories, ideas, and concepts continue to hold after they've come in continued contact with the world. Self-citation and self-quoting says, "Hey, this still works!" because so often it doesn't. I talk to many young researchers who are worried about setting their thoughts to paper because they might later change their minds. I hope you do! You will never get it *right* or *done* if you are thinking and growing. Publishing marks where you are on that path at that moment. Self-citing extends that path.

22 Missing and Murdered Indigenous Women and Girls.

low a threshold from which the rivers could recover.[23] But his theory of self-purification-cum-assimilative-capacity also assumed access to Indigenous Land. Phelps not only accessed Indigenous Land along the Ohio River to do his science; he also routinized state access by advocating for all rivers on all lands to be governed—carefully! precisely!—as proper sinks for pollution. Whether motivated by profit and growth or environmental conservation, both approaches to waste and wasting are premised on an assumed entitlement to Indigenous Land.

That's colonialism.

While there are different types of colonialism—settler colonialism, extractive colonialism, internal colonialism, external colonialism, neoimperialism—they have some things in common. Colonialism is a way to describe relationships characterized by conquest and genocide that grant colonialists and settlers "ongoing state access to land and resources that contradictorily provide the material and spiritual sustenance of Indigenous societies on the one hand, and the foundation of colonial state-formation, settlement, and capitalist development on the other."[24] Colonialism is more than the intent, identities, heritages, and values of settlers and their ancestors. It's about genocide and access.[25]

Emphasizing the role of access to Indigenous Land for colonialism, Edward Said (Palestinian)[26] writes:

> To think about distant places, to colonize them, to populate or depopulate them: all of this occurs on, about, or because of land. The actual

23 Tarr, "Industrial Wastes and Public Health," 1060. Also see Phelps's own words in Phelps, "Discussion."

24 Coulthard, *Red Skin, White Masks*, 7.

25 In her important work bringing Indigenous studies and Black studies together in *The Black Shoals*, Tiffany Lethabo King makes a strong case that analytical frames originating in White settler colonial studies that foreground land, rather than genocide and conquest, as the defining feature of colonialism miss intersectionality and grounds for coalition politics between Black and Indigenous peoples. She writes, "Genocide—and the making of the Native body as less than human, or flesh—remains the focus and distinguishing feature of settler colonialism," and that "an actual discussion of Native genocide is displaced by a focus on White settlers' relationship to land rather than their parasitic and genocidal relationship to Indigenous and Black peoples" (56, 68). Yes, yes, yes. I also think that Land relations, and thus the emplacement of more-than-human relations, are one of the keystones to doing anticolonial work as a Métis scientist. So I focus on Land here, and the inheritance of scientific land relations, knowing that this is shorthand for genocide. Also see Trask, *From a Native Daughter*; and Trask, "The Color of Violence."

26 This self-identification is in Said, "Zionism from the Standpoint of Its Victims."

> geographical possession of land is what empire in the final analysis is all about. At the moment when a coincidence occurs between real control and power, the idea of what a given place was (could be, might become), and an actual place—at that moment the struggle for empire is launched. This coincidence is the logic both for Westerners taking possession of land and, during decolonization, for resisting natives reclaiming it.[27]

Let's take a moment to focus on that bit about Westerners. Western culture—the heritage of social norms, beliefs, ethical values, political systems, epistemologies, technologies, and legal structures and traditions heavily influenced by various forms of Christianity and Judaism that have some origin in Ancient Greece and which heavily influenced societies in Europe and beyond—is not synonymous with colonialism. Western culture certainly has its imperialistic and colonial impulses, histories, and ideas of what is good and right, but these are different things from colonialism. When I hear a researcher ask, "Isn't doing research ethics paperwork colonial?," they are conflating Western and colonial. Remember: treaties are paperwork. If paperwork is used to possess land and secure settler and colonial futures, then, yes, it's colonial. But there is also anticolonial, Western-style paperwork that accomplishes the opposite, like the forms required by Indigenous research ethics boards. Colonialism, first, foremost, and always, is about *Land*, including the circumvention of ethics paperwork so researchers can have unfettered and unaccountable access to field sites (a.k.a. homelands), archives, samples, and data.[28]

The focus on Land—what it could be, what it might become, what it is for—does not always mean accessing Land as property for settlement, though it often does. It can also mean access to Land-based cultural designs and culturally appropriated symbols for fashion. It can mean access to Indigenous Land for scientific research. It can mean using Land as a Resource, a practice that may generate pollution through pipelines, landfills, and recycling plants, or as a sink to store or process waste. It can mean imagining a clean, healthy, and pollution-free future and conducting beach cleanups on Indigenous Land without permission or consent. It means imagining things for land in ways that align with colonial and settler goals, even when those goals are well intentioned. Especially when they are well intentioned. Which means it's time to talk about environmentalism.

27 Said, *Culture and Imperialism*, 93.

28 E.g., Lawford and Coburn, "Research, Ethnic Fraud, and the Academy."

Environmentalism and Colonialism

Environmentalism does not usually address colonialism and often reproduces it. Philosopher Kyle Whyte (Potawatomi),[29] Dina Gilio-Whitaker (Colville Confederated Tribes),[30] and many others[31] have pointed out that environmental solutions to pollution such as hydroelectric dams,[32] consumer responsibility, and appeals to the commons[33] assume access to Indigenous Land and its ability to produce value for settler and colonial desires and futures. Environmentalism often "propagate[s] and maintain[s] the dispossession of [I]ndigenous peoples for the common good of the world."[34]

For example, in September 2015, a US-based environmental NGO called the Ocean Conservancy released a report looking for solutions to marine plastic pollution that recommended that countries in Southeast Asia work with foreign-funded industries to build incinerators to burn plastic waste.[35] This recommendation follows a long line of colonial acts in the name of plastics, from accessing Indigenous Land to extracting oil and gas (and occasionally corn) for feedstock; to producing disposable plastics that use land to store, contain, and assimilate the waste; to pointing the finger at local "foreign" and Indigenous peoples for "mismanaging" waste imported from industrial and colonial centres; and then gaining access to that Land to solve their uncivilized approach to waste (mis)management.[36]

This is not to say that the Ocean Conservancy is evil, or even aware of its colonial mindset. Colonialism doesn't come from asshat goons, though it cer-

29 Whyte, "Dakota Access Pipeline."

30 Gilio-Whitaker, *As Long as Grass Grows.*

31 paperson, "Ghetto Land Pedagogy"; Osborne, "Fixing Carbon, Losing Ground"; Osborne, Bellante, and vonHedemann, *Indigenous Peoples and REDD+.*

32 Nunatsiavut Government, "Make Muskrat Right."

33 Fortier, *Unsettling the Commons.*

34 Byrd, *Transit of Empire*, xix.

35 Ocean Conservancy, "Stemming the Tide."

36 The term *mismanaged waste* has gained traction since a scientific publication estimating the amount of plastics entering the oceans used the category of mismanaged waste to estimate plastic leakage from land to the ocean. The problem is that everyone whose waste management did not look like the United States was automatically labelled *mismanaged.* The term signals that the infrastructure in question isn't quite Civilized enough. A detailed critique of this study and its colonial premises is in chapters 1 and 2. For community and grassroots pushback to this report, see GAIA Coalition, "Open Letter to Ocean Conservancy."

tainly has a large share of such agents. Colonial land relations are inherited as common sense, even as good ideas.[37] Many environmental historians have shifted their understanding of the origins of environmentalism well before back-to-the-land and save-the-(access-to-)land movements of the 1960s and 1970s. Instead they highlight earlier imperial archiving, cultivation, and control measures necessary for the flourishing of empire around the globe, both within and outside of what is lately called North America.[38] They argue that the colonial scientists who attempted to mitigate and halt environmental destruction in colonies so that the colonies might flourish are "the pioneers of modern environmentalism,"[39] where "environmentalism is police action, inseparable from western conceptions and attitudes"[40] of how to best organize and govern land (more on this in chapter 1).

The way that environmental crises and their solutions maintain rather than change existing power structures is central to the scholarship of anthropologist Joseph Masco (settler), who points out that "crisis," environmental and otherwise, has "become a counterrevolutionary idiom in the twenty-first century, a means of stabilizing an existing condition rather than minimizing forms of violence across militarisms, economy, and the environment."[41] Rather than using crisis as a relational model that puts certain things beyond dispute in the imperative to act at all costs, I focus on colonial land relations within environmental narratives and action as a way to acknowledge and address this usually unmarked power dynamic.

37 Here, I am drawing on Foucault's (unmarked) articulation of power as regimes of truth that allow some things to make sense, to circulate, and to act as truth, while others do not. See Foucault, *Discipline and Punish*. However, following Michelle Murphy (Métis), I build on this work "unfaithfully," as "Foucault's own work on neoliberal economics refuses to engage with colonial and postcolonial histories, the elaboration of the racial state, and drops sex as a central analytic." Murphy, *Economization of Life*, 149.

38 Anker, *Imperial Ecology*; Komeie, "Colonial Environmentalism."

39 Grove, "Origins of Environmentalism," 12. I think Grove and I see eye to eye on the term *pioneer* here.

40 Barton, *Empire Forestry*, 6.

41 Masco, "Crisis in Crisis," S65. Also see Masco, "Bad Weather." Joe Masco, thank you not only for your excellent, careful, original, and insightful work on the links between environmental and military crises, but, more importantly (to me and as a model in the academy), for your genuine generosity, solid and obvious forms of support, forceful and inspiring yet gentle curiosity, and feminist, caring ways that you invest in emerging intellectuals. Thank you, Joe, for taking time and care to be part of this book's life (and mine!).

Capitalism and Colonialism

To change colonial land relations and enact other types of Land relations requires specificity. This is so we don't accidentally think that the opposite of colonialism is environmentalism or, similarly, that we don't conflate colonialism with other forms of extraction, such as capitalism. Colonialism and capitalism might be happy bedfellows and indeed longtime lovers, but they are not the same thing.

Political economist Karl Marx (unmarked) argues that primitive accumulation (the stealing of land) is foundational to the possibility of capitalism—it's how someone gets more capital than someone else in the first place, which you need to jump-start a system where only a few people own the means of production.[42] You can't make and hoard capital without stealing Land first. We have case studies of how aspects of capitalist production and technologies allow specific forms of colonialism and dispossession to take root and spread.[43] Likewise, excellent research describes the sweet trifecta of capitalism, colonialism, and pollution. The treadmill of industrial and capitalist production is ever in need of more Land to contain its pollution,[44] leading to the argument that "contamination and resource dispossession [are] necessary and inherent factors of capitalism."[45]

Yet colonial quests for Land are different than capitalist goals for capital, even if pollution has a role in attaining each goal. Socioeconomic systems other than capitalism also create environmental pollution and waste,[46] but what is more important for understanding the relationship between capitalism and colonialism is that many different economic systems depend on access to Indigenous Land. As Sandy Grande (Quechua) has argued, "Both Marxists and capitalists view land and natural resources as commodities to be exploited, in the first instance, by capitalists for personal gain, and in the second by Marxists for the good of all."[47] Eve Tuck (Unangax) and Wayne Yang (diaspora settler of colour) have pointed out, "Socialist and communist empires have also been settler empires (e.g., Chinese colonialism in Tibet)."[48] Colonialism is not one kind of

42 Marx, "The Modern Theory of Colonisation," chap. 33 in *Capital*, vol. 1.

43 Denoon, *Settler Capitalism*; Pasternak, "How Capitalism Will Save Colonialism."

44 Voyles, *Wastelanding*.

45 Ofrias, "Invisible Harms, Invisible Profits," 436.

46 Gille, *From the Cult of Waste*; Kao, "City Recycled"; Scheinberg and Mol, "Multiple Modernities." We need a lot more research in this area.

47 Grande, *Red Pedagogy*, 31.

48 Tuck and Yang, "Decolonization Is Not a Metaphor," 4.

thing with one set of techniques that always align with capitalism. Marxism, socialism, anticapitalism, capitalism, and other economic systems can, though certainly don't have to, enact colonial relations to Land as a usable Resource that produces value for settler and colonizer goals, regardless of how and by whom that value is produced.

Colonialism, capitalism, and environmentalism do not have settled relationships or forms.[49] For instance, colonialist states and powers have at times sided with environmental conservation over capitalist gains. Historians have documented how, as Richard Grove (unmarked) puts it, "Paradoxically, the colonial state in its pioneering conservationist role provided a forum for *controls* on the unhindered operations of capital for short-term gain which, it might be argued, brought about a contradiction to what is normally supposed to have made up the common currency of imperial expansion. Ultimately, the long-term security of the state, which any ecological crisis threatened to undermine, counted for far more than the interests of private capital bent on the destruction of the environment."[50] To make capitalism and colonialism synonymous, or to conflate environmentalism and anticolonialism, misses these complex relations.

Because of this nuance and its repercussions for political action, political scientist Glen Coulthard (Yellowknives Dene) has called for scholars to shift their analysis away from capitalist relations (production, proletarianization) to colonial relations (dispossession, Land acquisition, access to Land): "Like capital, colonialism, as a structure of domination predicated on dispossession, is not a 'thing,' but rather the sum effect of the diversity of interlocking oppressive social relations that constitute it. When stated this way, it should be clear that shifting our position to highlight the ongoing effects of colonial dispossession in no way displaces questions of distributive justice or class struggle; rather, it simply situates these questions more firmly alongside and in relation to the other sites and relations of power that inform our settler-colonial present."[51] Conflating colonialism with capitalism misses crucial relations, which Coulthard argues include white supremacy and patriarchy. Aileen Moreton-Robinson (Geonpul,

49 Feminist geographers like J. K. Gibson-Graham (unmarked) have done excellent work showing how capitalism is not only diverse in its manifestations, but also patchy and incomplete. They argue that to describe capitalism as a total and complete system is to give it power it does not necessarily have. Gibson-Graham, "End of Capitalism"; Gibson-Graham, "Rethinking the Economy."

50 Grove, "Origins of Environmentalism," 12; emphasis added. This is an appropriate use of the term *pioneering*.

51 Coulthard, *Red Skins, White Masks*, 15.

Quandamooka First Nation) has shown that it misses racial formations and racism.[52] For thinkers such as Tuck and Yang, the "homogenization of various experiences of oppression as colonialism"—that is, conflating imperialism, racism, capitalism, exclusion, and general bad behaviour with colonialism—accomplishes "a form of enclosure, dangerous in how it domesticates decolonization. It is also a foreclosure, limiting in how it recapitulates dominant theories of social change."[53]

Differentiation and specificity matter to ensure that actions address problems, and the conflation of colonialism with other ills ensures the erasure of horizons of meaningful action that can attend specifically to assumed settler and colonial entitlement and access to Land. In the case of pollution, a focus on capitalism misses relations that make Land available for pollution in the first place. It can miss the necessary place of stolen Land in colonizers' and settlers' ability to create sinks for pollution *as well as* stolen Land's place in alternative economies (via a communal commons) and environmental conservation (via methylmercury-producing hydroelectric dams).

Pollution, scientific ways to know pollution, and actions to mitigate pollution are not examples of, symptoms or metaphors for, or unintentional byproducts of colonialism, but rather are essential parts of the interlocking logics (brain), mechanisms (hands and teeth), and structures (heart and bones)

52 Moreton-Robinson, *White Possessive*. Thank you, Aileen Moreton-Robinson, for the political and intellectual move of foregrounding identity and culture as the primary grounds from which to make claims and change. I think this is a key lesson for activism: "Patriarchal white nation-states and universities insist on producing cultural difference in order to manage the existence and claims of Indigenous people. In this way the production of knowledge about cultural specificity is complicit with state requirements for manageable forms of difference that are racially configured through whiteness." Moreton-Robinson, *White Possessive*, xvii.

53 Tuck and Yang, "Decolonization Is Not a Metaphor," 17, 3. I wish to express a deep gratitude for your work, Eve Tuck, and especially for "Suspending Damage," which has profoundly shaped my research, including the way this book was framed and written. Tuck's open letter is, in many ways, directly responsible for turning my work from being about plastic to being about colonialism. It is part of a shift that took place in my scientific work from attempting to create an accounting of chemical harms by counting plastic to articulating food sovereignty (details on this method are in chapter 3). I re-read "Suspending Damage" and "Decolonization Is Not a Metaphor" at least once a season, as an event to sit with the text, rather than as a source to pull things from (a reading technique I strengthened after reading some of your tweets on extractive reading practices). Your work has easily been some of the most formative in my intellectual and ethical journey. Thank you, Eve Tuck, for your brilliance, pedagogy, and ethics.

of colonialism that allow colonialism to produce and reproduce its effects in Canada, the United States, and beyond.[54] Colonialism is not just about taking Land, though it certainly includes taking Land. Stealing is a manifestation, a symptom, a mechanism, and even a goal of colonialism. But those are the teeth of colonialism, and I want to look at its bones. Stealing Land and dispossessing people are events with temporal edges, but ongoing Land theft requires maintenance and infrastructure[55] that are not as discrete, given that "colonization is a continuing process, not simply a historical event."[56] Colonialism is a set of specific, structured, interlocking, and overlapping relations that allow these events to occur, make sense, and even seem right (to some).[57] I will argue throughout this text that these relations—their types, durations, effects, and maintenance—are also enacted by pollution and pollution science.

Otherwises and Alterlives

When I first began researching plastic pollution around 2008, I thought that plastics had the immense potential to blow concepts of pollution out of the water,[58] since they defy so many scientific and popular truisms. You can't "clean up"

54 There are different colonialisms, imperialisms, and indigeneities because these things are place- and time-based. When I speak in general terms, statements are rooted in relations from Newfoundland and Labrador and early teachings in Alberta, Canada. They will not make global sense (more on the difference between universalism and generalization of knowledge in chapter 3).

55 For an example of interlocking infrastructures at multiple scales that maintain Land theft (even as they fail!), see Pasternak, *Grounded Authority*. This text is particularly good for discussions of how Indigenous jurisdiction and Land are consistently usurped in place, particularly by the state through mechanisms of financialization and "accountability." It is also an excellent text for studying/punching up, for showing how Canadian state sovereignty and jurisdiction consistently fall short and are patchy, even though they are often assumed to be solidly in place. Thank you, Shiri Pasternak (settler), for your excellent work.

56 Anguksuar, "Postcolonial Perspective." Also see the more oft-cited Wolfe, *Settler Colonialism*.

57 Sandy Grande writes about the animating beliefs and logics that underpin colonial societies that serve as the basis for common sense. These core beliefs are as follows: (1) belief in progress as change and change as progress; (2) belief in the effective separateness of faith and reason; (3) belief in the essential quality of the universe and of "reality" as impersonal, secular, material, mechanistic, and relativistic; (4) subscription to ontological individualism; and (5) belief in human beings as separate from and superior to the rest of nature. While this text focuses on the third and fifth beliefs, and particularly how they manifest in pollution science, all five are part of how land is understood and related to. Grande, *Red Pedagogy*, 69.

58 Pun!

plastics because they exist in geological time, and cleaning just shuffles them in space as they endure in time.[59] You can't recycle them out of the way, because it means ever more will be produced,[60] and there is no "away" at any rate.[61] Many of the chemicals associated with plastics, called endocrine disruptors, defy thresholds and exceed the adage that the "danger is in the dose" or the "solution to pollution is dilution" because they cause harm at trace quantities already present in the environment and bodies.[62] Plastics and their chemicals defy containment, a hallmark approach to industrial waste management, as they blow, flow, and off-gas so that their pollutants are ubiquitous in every environment tested.[63] Last but hardly least, their long temporality means their future effects are largely unknown,[64] making uncertain the guarantee of settler futures. I thought these traits would provide pollution science and activism with the case they needed to move beyond thresholds of allowable harm, beyond disposability, and beyond the access to Land that both thresholds and "away" require.[65] But despite con-

59 Gray-Cosgrove, Liboiron, and Lepawsky, "Challenges of Temporality."

60 MacBride, "Does Recycling Actually Conserve or Preserve Things?" Thank you, Samantha MacBride (unmarked). You are one of the smartest, most careful, most multiscalar and interdisciplinary thinkers I have had the pleasure to know intellectually (and personally!) when it comes to waste streams and recycling in the United States. You are a role model for how you put your intelligence to work as the director of research at the New York City Department of Sanitation. If I had to teach only one text on waste, it would be yours: MacBride, *Recycling Reconsidered*. Thank you, Samantha MacBride, for all the forms of work you do and particularly how you do it.

61 Davies, "Slow Violence and Toxic Geographies"; Bullard, *Dumping in Dixie*.

62 E.g., Vandenberg, "Low-Dose Effects of Hormones and Endocrine Disruptors."

63 Bergman et al., "Impact of Endocrine Disruption," A104; vom Saal et al., "Chapel Hill Bisphenol A Expert Panel Consensus Statement," 131.

64 You may have noticed that temporal estimates of plastics breaking down (one thousand years for this kind of plastic, ten thousand for this other kind) exceed the amount of time that plastics have existed. Most of these estimates are modeled from data created in labs (in UV-saturated, vibrating, acidic set-ups that rarely mimic actually existing environmental conditions) and are based on the idea that the rate of weakening polymer bonds will proceed on a regular curve. They do not anticipate the effects of metabolites or the molecular chains that polymers might break into. They cannot anticipate how future environmental relations will absorb, adapt to, and otherwise influence these rates of breakdown or the effects of many types of plastics in diverse environments over long periods.

65 This is what feminist STS scholars such as Martha Kenney (unmarked) and others might call *response-ability*: "*cultivating the capacity for response*. Recent works in feminist science studies have proposed *response-ability* as a term that might whet our imaginations for more relational ethics and politics enacted in everyday practices of living in our more-

siderable and sustained public, scientific, and policy attention to plastic pollution, most pollution science and activism have not shifted this way (with a few notable exceptions[66]).

As feminist scholar Susan Leigh Star (unmarked) reminds us, "It might have been otherwise."[67] In fact, it has been. There are and have been other definitions of and relations to pollution. Not all pollution is colonial, but the idea of modern environmental pollution[68] certainly is (more on this in chapter 1). Be-

---

than-human world." Kenney, "Fables of Response-Ability," 7; emphasis in original. Also see work by María Puig de la Bellacasa (unmarked), Donna Haraway (unmarked), Alexis Shotwell (unmarked), Karen Barad (unmarked), Lucy Suchman (unmarked), Kim Fortun (unmarked), Aryn Martin (unmarked), Natasha Myers (settler), Michelle Murphy (Métis), Shawn Wilson (Cree), Dwayne Donald (Cree), Zoe Todd (Métis), Kim TallBear (Sisseton-Wahpeton Oyate), Sara Tolbert (unmarked), and Winona LaDuke (Anishinaabe) on accountability and responsibility in relations.

66 Settler scientists such as Chelsea Rochman (unmarked), Laura Vandenberg (unmarked), and Fred vom Saal (unmarked), among others, have all written about the chemical hazards of plastics and their associated chemicals and the way science, industry, and policy ought to relate to one another. They work within dominant science to shift the conversation. I'll speak more about some of their work in chapter 2. See, e.g., Rochman et al., "Policy"; Vandenberg et al., "Regulatory Decisions on Endocrine Disrupting Chemicals"; vom Saal and Hughes, "Extensive New Literature." Global Alliance for Incinerator Alternatives (GAIA) is also exemplary for its insistence in looking upstream at industry and political alliances for the source of marine plastics and has folded critiques of capitalism and colonialism into its work. GAIA has also proposed some shifts in scientific methods of monitoring marine plastics, which I discuss in chapter 2. See GAIA, "Plastics Exposed."

67 Star, "Power, Technology, and the Phenomenology of Conventions," 53.

68 I use the term *modern pollution* to mean post-miasma theories of environmental pollution based on quantitative science, threshold limits, and industrial capture. In *Risk and Blame*, white primitivist anthropologist Mary Douglas (British) differentiates between cultural notions of pollution and "technical" senses: "There is a strict technical sense, as when we speak of river or air pollution, when the physical adulteration of an earlier state can be precisely measured. The technical sense rests upon a clear notion of the prepolluted condition. A river that flows over muddy ground may be always thick; but if that is taken as its natural state, it is not necessarily said to be polluted. The technical sense of pollution is not morally loaded but depends upon measures of change. The other sense of pollution is a contagious state, harmful, caused by outside intervention, but mysterious in its origins." Douglas, *Risk and Blame*, 36. But one of my primary arguments is that this "technical" sense of pollution is indeed morally loaded with the values and goals of colonialism and that there is therefore no real difference between Douglas's categories. I nevertheless use the term *modern environmental pollution* to highlight, as Douglas does, the recent origins and culturally specific aspects of scientific definitions of pollution.

fore the threshold model of pollution pioneered[69] by Streeter and Phelps, there were many definitions of pollution that shared a more prohibitive and normative slant. The English word *pollution* comes from the Latin *pollutionem*, meaning defilement or desecration. The earliest recorded uses in the mid-fourteenth century refer to the "discharge of semen other than during sex."[70] This may seem like a brilliant idea, but in the Christian Middle Ages extracoital dissemination was written up as an act of desecration, an interruption of the true and right path for semen. Pollution was (and still is) about naming a deviation from the good and true path of things—good relations manifested in the material. Though it wasn't until 1860 that the term *pollution* was recorded in the sense of environmental contamination,[71] the morality and ideas of good and right paths for contaminants remain a key aspect of understanding pollution today. These moral overtones still circulate in environmental science even while we scientists argue that we are measuring wayward particles rather than immoral acts.[72]

Both pollution and plastics have been otherwise, with different and varied interpretations and enactments. The stakes of my research are to open up plastics and pollution so that they are otherwise, yet related, once more (and still). By denaturalizing and demythologizing pollution in general and plastics in particular, I aim to make (more) apparent their ongoing relationships to maintaining colonial Land relations as well as to anticolonial Land relations. That way, when we want to do scientific and/or activist work that does not reproduce colonial L/land relations, we know where we stand and what we mean.

69 Yes, pioneered in the spirit of land acquisition via frontierism and the erasure of other forms of Land relation.

70 Online Etymology Dictionary, s.v. "pollution," accessed August 12, 2020, https://www.etymonline.com/word/pollution.

71 Online Etymology Dictionary, s.v. "pollution."

72 An interesting example of this is that environmental scientists consistently eschew their training to say that the presence of plastics in environments is a form of harm, while the dominant scientific model of pollution distinguishes between contamination (presence) and pollution (demonstrated harm). In "The Ecological Impacts of Marine Debris," Chelsea Rochman and collaborators argue that conflating the two might actually work against conservationist goals, since it gives a space for the plastics and petrochemical industries to defer action by saying harm must be demonstrated beyond presence. I agree with Rochman et al. in a sense. But I extend their argument to say that embracing an idea of pollution as bad relations that can exceed scientific evidence of harm is exactly what we need. If you're going to go with a more overtly "anthropological" set of value-based definitions of pollution as bad relations, do it and do it loud, which means not conflating it with other (scientific) models of pollution with different values and goals.

As such, my orientation for this book is a specific enactment of a *particular* otherwise. Following Michelle Murphy's concept of alterlife, I seek "words, protocols, and methods that might honor the inseparability of bodies and land, and at the same time grapple with the expansive chemical relations of settler colonialism that entangle life forms in each other's accumulations, conditions, possibilities, and miseries."[73] When I am taking plastics out of birds' gizzards one by one with tweezers, I am searching for these words, protocols, and methods *as a scientist*. I want to know whether or how to use an available threshold-based measurement in plastic pollution research (called the EcoQO) when I don't think threshold models are in good relation yet know that the measurement is one of the few effective for policy. I think about how my colleague got this bird to begin with—was it in good relations, or did it assume entitlement to Land? Whose water am I using to clean these plastics, anyhow? And, most importantly, when Murphy writes, "The concept of alterlife is offered as a way of approaching the politics of relations in solidarity with the vast labor of antiracist and decolonial reproductive and environmental justice activism, as well as Indigenous survivance and resurgence,"[74] the methodological question is: how do I get to a place where these relations are properly scientific, rather than questions that fall outside of science, the same way ethics sections are tacked on at the end of a science textbook? How do I, as a scientist, make alterlives and good Land relations integral to dominant scientific practice?

There is no terra nullius for this work. Western science has long been identified as a practice that assumes mastery over Nature, reproduces the doctrine of discovery, revels in exploration and appropriation of Indigenous Land, and is invested in a rigorous self-portraiture[75] in which valid scientific knowledge is created only by proper European subjects.[76] It's also pretty sexist. But dominant science[77] is my terrain. At CLEAR, we use science against science, understand-

73 Murphy, "Alterlife and Decolonial Chemical Relations," 497. Thank you, Michelle Murphy, for so many reasons. For your scholarship, which has grounded the thinking of multiple generations of STS scholars, and for the way you mentor and create spaces, lessons, and examples for good relations in academia and beyond. Your work and practices make diverse futures for so many of us (a.k.a. legacy). I cannot overstate the effects of your intelligence, generosity, and ethics on me and so many others. Maarsi.

74 Murphy, "Against Population, towards Alterlife," 118.

75 Daston, "History of Science."

76 Seth, "Putting Knowledge in Its Place."

77 I use the term *dominant science* instead of *Western science* for two reasons. First, *dominant* keeps the power relations front and centre, and it's these power relations I am usually discussing. Western science is a cultural tradition where ways of knowing start with the

ing that science is always already fucked up, which means that our work is always compromised (a concept I explain more in chapter 3). To imagine a clean slate from which to start our anticolonial science is to subscribe to "terra nullius, the colonizer's dream," described by feminist scholar Raewyn Connell (settler) as "a sinister presupposition for social science. It is invoked every time we try to theorise the formation of social institutions and systems from scratch, in a blank space. Whenever we see the words 'building block' in a treatise of social theory, we should be asking who used to occupy the land."[78] Research and change-making, scientific or otherwise, are always caught up in the contradictions, injustices, and structures that already exist, that we have already identified as violent and in need of change.[79] This text is about maneuvering within this complex and compromised terrain.

This compromise of doing both Indigenous and anticolonial work in science and academia[80] is something that many Indigenous thinkers contend with when they enter academia.[81] CLEAR member Edward Allen (Kablunangajuk) opens his doctoral comprehensive exam with the following words:

> The academy will have to embrace wholesale change in what it qualifies as legitimate knowledge production and pedagogy if it is to capture any Indigenous knowledges in any meaningful way.[82] Until the hurdles are cleared, I will continue to write as if footed in both worlds. This with op-

---

Ancient Greeks, get influenced by various forms of Christianity and Judaism, and move through the Enlightenment. Generally, I have no problem with that culture. The problem is when it becomes dominant to the point that other ways of knowing, doing, and being are deemed illegitimate or are erased. Second, not all Western science is dominant. Midwifery, alchemy, and preventative medicine are part of Western science that suffer at the hands of dominant science.

78 Connell, *Southern Theory*, 46.

79 For an excellent example of how the politics of denunciation can reproduce the wider system of uneven power relations that it seeks to denounce, see Fiske, "Dirty Hands." For more on what is compromised in conducting basic science for justice, including community science, see Shapiro, Zakariya, and Roberts, "Wary Alliance." For more on how many scientists already know this, see O'Brien, "Being a Scientist."

80 Many academics state that academia is colonial, and they're quite right. But they usually aren't specific as to the intentional roles that universities played in imperialism and the disciplining and oppressions of Indigenous peoples. Now you can be specific: Pietsch, *Empire of Scholars*. But you can also be nuanced and generous: paperson, *A Third University Is Possible*.

81 E.g., S. Wilson, *Research Is Ceremony*; A. Simpson, "On Ethnographic Refusal."

82 He cites Bang, Medin, and Cajete, "Improving Science Education for Native Students."

> timism of at least some small piece of the original story being heard, to imitate my Elders (and my occasional Western teacher) who speak from the heart and exercise compassion when faced with shortcomings (as has been done repeatedly for me), and to reluctantly trade the risk of harm for any opportunity to contribute to change from the inside. But, in the short list of things I claim to grasp, I am confident that you *cannot* come to a full understanding of Indigenous concepts of relationality in this [written] format, even if I were to produce here the best academic paper ever written.[83]

These existing terrains are the fertile, toxic grounds[84] for alterlife:

> A politics of non-deferral that is a commitment to act now. But this politics of non-deferral is not driven by the logic of the emergency, the scale of the planetary, or the container of the nation state. It is a politics of non-deferral interested in the humbleness of right here, in the scale of communities, and in the intimacies of relation. Alterlife is a challenge to invent, revive, and sustain decolonizing possibilities and persistences right now as we are, forged in non-innocence, learning from and in collaboration with past and present projects of residence and resurgence.[85]
>
> Let's begin.

Differences and Obligations

Different groups have different roles in alterlives, reconciliation, decolonization, indigenization, and anticolonial work. An ongoing issue at CLEAR, which includes Indigenous people, local and come-from-away settlers, as well as those who are neither Indigenous nor settler, such as international students from Nigeria,[86] is how to take up science that enacts good Land relations without appropriating Indigenous Land relations if they aren't yours (including when they belong to a different Indigenous group). I keep talking about specificity. Here, I think of specificity as a methodology of nuanced connection and humility,[87]

83 E. Allen, "Neighboring Ontologies."

84 Land can be polluted and still foster good land relations. See, e.g., Konsmo and Recollet, afterword; and Hoover, "Cultural and Health Implications of Fish Advisories," 4.

85 Murphy, "Against Population, towards Alterlife," 122–23.

86 Vowel, *Indigenous Writes*.

87 For more on humility, see L. Simpson, *Dancing on Our Turtle's Back*; and Kimmerer, *Braiding Sweetgrass*.

rather than as a way to substantiate uniqueness. Anthropologist Tim Choy's (unmarked) work is exemplary for showing how specificity, when used methodologically, has varied political allegiances and outcomes, from speciesism to state autonomy.[88] Rather than mobilize specificity and particularism for categorization, I want to call attention to their ability to situate differences that matter to political action.[89]

Problems, Theories, and Methods of We

The joke was old even before it appeared in print:

> The Lone Ranger and Tonto find themselves surrounded by hostile Indians. The Ranger asks Tonto: "What are we going to do, Tonto?" To which Tonto replies: "What do you mean we, white man (or paleface, or kemo sabe, depending on the version)?"
>
> Its racist ancestry is undeniable: the joke partly evokes the picture of a feckless subordinate who will treacherously abandon his superior at the first sign of trouble—usually with the ethnic or social group to which the subordinate belongs. But even before 1956, ancient variants of the joke were meant to deflate the condescension of individuals who used the royal "we," and the insulting presumption of people who assumed, for their own purposes, what they had no business assuming.[90]

We is rife with such assumptions. A familiar, naturalized narrative about environmental pollution is that We are causing it. We are trashing the planet. Humans are inherently greedy, or wasteful, or addicted to convenience, or naturally self-maximizing, and are downright tragic when it comes to "the" commons. On the other side of the coin, We must rise up, work together, refuse plastic straws, act collectively, and put aside our differences.

I'm not going to dwell on how We erases difference and power relations, or how it makes a glossy theory of change that doesn't allow specific responsibil-

88 Choy, *Ecologies of Comparison*.

89 This is what feminist Elizabeth Grosz (unmarked) might define as the type of difference that is "not seen as different from a pregiven norm, but as pure difference, difference in itself, difference with no identity." Grosz, "Conclusion," 339.

90 Ivie, "What Do You Mean 'We,' White Man?" Also see Heglar, "Climate Change Ain't the First Existential Threat"; Hecht, "African Anthropocene"; and Whyte, "Is It Colonial Déjà Vu?" All of these pieces break out of the violence and myopia of "we" as a way to critique mainstream environmental narratives, including the notion of the Anthropocene (which is also a key critique in Murphy, "Alterlife and Decolonial Chemical Relations").

ity.[91] Here, I want to focus on responsibility—the obligation to enact good relations as scientists, scholars, readers, and to account for our relations when they are not good. And you can't have obligation without specificity.

We isn't specific enough for obligation. You know this—an elder daughter has different obligations than a mail carrier, and you have different obligations to your elder daughter than to the mail carrier. DuPont has different obligations to plastic pollution than someone with a disability who uses a straw to drink. Even though I'm sure you've heard that "everything is related" in many Indigenous cosmologies, this doesn't mean there is a cosmic similitude of relations. You are not obliged to all things the same way.[92] Hence there is a need for specificity when talking about relations.

There can be solidarity without a We. There *must* be solidarity without a universal We. The absence of We and the acknowledgement of many we's (including those to which you/I/we do not belong[93]) is imperative for good re-

91 If you want some more of that, see M. Liboiron, "Against Awareness, for Scale"; and M. Liboiron, "Solutions to Waste." There is also an entire chapter on the problems of We in a currently in-progress manuscript called *Discard Studies* that I am writing with excellent collaborator Josh Lepawsky (settler).

92 The idea that obligations are specific is put into practice by many different Indigenous thinkers, but this guiding principle is not exclusive to Indigenous groups. I think of New Orleans activist Shannon Dosemagen (unmarked), director of the Public Lab for Open Technology and Science, whose understandings of relations as the primary source, goal, and ethic of community science have led to a career in bringing people together in a good way and building technologies and platforms to support those relations. See Dosemagen, Warren, and Wylie, "Grassroots Mapping." I also think about Labrador-based scholar Ashlee Cunsolo (settler), director of the Labrador Institute, whose directorship is premised on building and maintaining relations in a context of complex geopolitics and competing interests, and who exemplifies humility, generosity, and gratitude in every setting I've seen her in. See Cunsolo and Landman, *Mourning Nature*. Shannon and Ashlee, thank you for your examples of putting the relational politics that so many people talk about into practice in ways that far exceed the cultural and ethical norms of your existing institutions. It has been a great gift being activist-administrators with you.

93 Acknowledging where you do not belong while remaining aligned with those who do seems to be one of the more difficult lessons of allyship. I recently attended an "Indigenous LGBTQ2S+" gathering where white and non-Indigenous allies were thanked for attending, but then asked to leave so we could build a certain type of community. The settler sitting beside me didn't leave. She was clearly nervous and unsure of what to do, but her inability to choose the embarrassment of standing up and leaving, and thereby outing herself as a white person, over the choice to stay in a place she had been asked to leave by those she was there to support meant that she probably isn't ready for the even harder choices involved in allyship. Because of her choice to stay, I have never been in a room filled only

lations in solidarity against ongoing colonialism and allows cooperation with the incommensurabilities of different worlds, values, and obligations. There are guidebooks to doing careful, specific solidarity work across difference.[94]

Indigenous science and technology studies (STS) scholar Kim TallBear has written about "standing with" as a methodological approach to doing research in good relation. In her work, she writes that she "had to find a way to study bioscientists (whose work has profound implications for indigenous peoples) in a way in which I could stand more within their community," rather than critiquing them from a place of confrontation and not-caring—an approach that she argues is bad feminist practice. She now moves "towards faithful knowledges, towards co-constituting my own knowledge in concert with the acts and claims of those who I inquire among."[95] Indigenous peoples, settlers, and others have different roles and responsibilities in the "challenge to invent, revive, and sustain decolonizing possibilities and persistences."[96] Rather than fixing or saving one another, "giving back,"[97] or assuming that ongoing colonial Land relations only harm Indigenous people, "within the condition of alterlife the potential for political kinship and alter-relations comes out of the recognition of connected, though profoundly uneven and often complicit, imbrications in the systems that distribute violence."[98] This is investment without assumed access to our subjects and areas of research.

---

with Indigenous queer folk. Because of her choice, I had to take time to teach her when she was ignorant of something a speaker said. You can stand with a group without standing in their midst. In fact, sometimes standing-with-but-over-there is the best place to stand. A similar story is told by Sara Ahmed in the context of trying to have a Black Caucus professional meeting in *On Being Included*. I'm sure you have your own stories.

94 Land, *Decolonizing Solidarity*; Gaztambide-Fernández, "Decolonization and the Pedagogy of Solidarity"; Walia, "Decolonizing Together"; TallBear, "Standing with and Speaking as Faith"; Amadahy and Lawrence, "Indigenous Peoples and Black People in Canada."

95 TallBear, "Standing with and Speaking as Faith," 5. Thank you, Kim, for your big, bold, out-in-public work and thinking as well as your tableside, quieter talks. I'm sure you know that your work—written scholarship, Twitter essays and jokes, gathering and organizing—props the door open for so many others, and for this I am grateful. Also, love the hair. Maarsi, Kim.

96 Murphy, "Against Population, towards Alterlife," 122–23.

97 TallBear writes about Gautam Bhan's (Indian) notion of "continuous and multiple engagements with communities and sites of research rather than a frame of giving back," which maintains a benevolent narrative of wealth and deficit. TallBear, "Standing with and Speaking as Faith," 2.

98 Murphy, "Against Population, towards Alterlife," 120.

Decolonization and Anticolonialism

These politics are why we call CLEAR an anticolonial lab rather than a decolonial lab. I follow collaborators Tuck and Yang when they argue that "decolonization doesn't have a synonym."[99] They write that decolonization means "repatriating land to sovereign Native tribes and nations, abolition of slavery in its contemporary forms, and the dismantling of the imperial metropole. . . . Decolonization is not equivocal to other anti-colonial struggles."[100] It means other things, too, since there are many colonizations and thus many decolonizations, but my dedication to this meaning comes largely from being an academic, where the verb *decolonize* is frequently invoked as something that you do to university courses, syllabi,[101] panels, and other academic nouns.[102] Yet in the face of all this "decolonization," colonial Land relations remain securely in place. Appropriating terms of Indigenous survivance and resurgence, like decolonization, is colonial. If we've been working together in this text up until now, I hope you can see the relationship of such a promiscuous use of *decolonization* with the definitions of colonialism above: it means settler and colonial access to Indigenous Land, concepts (like decolonization and indigenization), and lifeworlds to advance settler and colonial goals, even if they are benevolent ones. Especially benevolent ones. Probably not what is intended.

99 Tuck and Yang, "Decolonization Is Not a Metaphor," 3.

100 Tuck and Yang, "Decolonization Is Not a Metaphor," 31. There is a tradition where decolonization refers specifically to knowledge, and this tradition comes largely out of Latin America and parts of Africa. While those theories and activisms are crucial to where they come from, so, too, is a definition of colonialism that gives up no ground, here in occupied territory. I do not think that Indigenous theorists from either tradition are interested in the conflation and the erasure and de-placing of our/their respective struggles.

101 Zara, "I don't know who needs to hear this right now . . ."

102 In short, I believe this land-based definition of decolonization matters in spaces where land relations are not already a guiding orientation. There are many spaces where a hard line on definitions of decolonization may not be appropriate, given the diversity of Indigenous groups, colonized groups, and their decolonization efforts. But this is an academic text with mostly academic readers and as such I'll assume a good chunk of white and settler readers (hello!). I have watched Indigenous people doing a diversity of Indigenous science and even decolonial science, and then watched well-intentioned settlers appropriating those terms to describe their own activities and goals over and over and over. While I think academia is increasingly seeking to put land relations at the forefront of critique and theory, we're not good at carrying that commitment into action. So, I start here with the 101 and some edges on the sandbox.

This also means CLEAR, as a lab, does not *claim* to do I[illegible] Indigenous science refers to science done by and for Indigeno[illegible] Indigenous cosmologies. Botanist Robin Wall Kimmerer's (Potaw[illegible] *ing Sweetgrass*, where she narrates botany through Potawatomi tra[illegible] teachings, is an example of doing Indigenous science in academia.[103] [illegible] most Indigenous science is done outside of academia and we will nev[illegible] about it.) While some Indigenous members of CLEAR certainly engage i[illegible] In-digenous science the way Kimmerer does, it isn't available to all lab members nor should it be. Likewise, CLEAR's Indigenous lab membership also engages in decolonization based on diverse understandings and reclamations of Land relations, but this also isn't available to all CLEAR members or to all readers. Indigenous peoples do, use, and refuse Western and Indigenous sciences along a rich spectrum, but CLEAR is not primarily an Indigenous science lab.

As director of CLEAR, I identify our space as an anticolonial lab, where anticolonial methods in science are characterized by how they do not reproduce settler and colonial entitlement to Land and Indigenous cultures, concepts, knowledges (including Traditional Knowledge), and lifeworlds. An anticolonial lab does not foreground settler and colonial goals. There are many ways to do anticolonial science: in addition to Indigenous sciences, there are, for example, also queer, feminist, Afro-futurist, and spiritual land relations that are anticolonial. *Anticolonial* here is meant to describe the diversity of work, positionalities, and obligations that let us "stand with" one another as we pursue good land relations, broadly defined.

Plastics' Specificity

Let's bring the idea of specificity and obligation into plastics. The term *plastic* refers to many types of polymers with many, many associated industrial chemicals. Plastic pollution scientist Chelsea Rochman and colleagues have written about how treating all plastics as one type of thing has led "to simplified studies and protocols that may be inadequate to inform us of the sources and fate of microplastics, as well as their biological and ecological implications."[104] Plastic in the singular misses things that are rather central to plastic activism, plastic science, plastic policy, and other plastic relations. For example, the term *single-use plastics* includes medical plastics, disposable packaging, and other items. Conflating them can cause harm, particularly when there are calls to ban all single-use plas-

103 Kimmerer, *Braiding Sweetgrass*.
104 Rochman et al., "Rethinking Microplastics as a Diverse Contaminant Suite," 703.

...he #suckitableism movement and thinker-advocates such as Alice Wong (unmarked) have been very clear that plastic bendy straws are used by people with disabilities to create livable worlds and that bans are ableist.[105] Without differentiating between medical plastics[106] (while also making them less toxic, as Health Care without Harm[107] is advocating) and other single-use plastics, or differentiating between PVC (which is full of toxic chemicals) and silicone (less so),[108] or differentiating between plastic use and plastic production, it is impossible to be responsible to the problems and ethics of plastic pollution (see chapter 2). This is just one way to think about the relationships among differentiation, specificity, ethics, and obligation in plastics.[109] There's not even a We for plastics.[110]

This Text Has Relations and Obligations

This text has specific obligations and relations as well. It was written on Beothuk Land in St. John's in the province of Newfoundland and Labrador: "The relationship between an object and where it belongs is not simply fortuitous, or a matter of causal forces, but it is rather intrinsic or internal, a matter of what that thing actually is."[111] Things like this book. Things like ideas. Place-based

105 Wong, "Rise and Fall of the Plastic Straw."

106 Jody Roberts (unmarked) has written about this issue eloquently in "Reflections of an Unrepentant Plastiphobe": his fear and dislike of plastics confront the medical plastics that keep his daughter alive. His work highlights how ethics and obligation are situated.

107 Health Care without Harm, "Health Care without Harm."

108 This is one of my points in M. Liboiron, "Redefining Pollution and Action."

109 A lot of social science work on plastics aims to denaturalize the social singularity of plastics. Most of this work attends to the minutia of the circulation, representation, re/use, or materiality of plastics in-place. For example, see H. Davis, "Life and Death in the Anthropocene"; H. Davis, "Toxic Progeny"; H. Davis, "Imperceptibility and Accumulation"; De Loughry, "Petromodernity"; De Loughry, "Polymeric Chains and Petrolic Imaginaries"; De Wolff, "Plastic Naturecultures"; De Wolff, "Gyre Plastic"; Gill, *Of Poverty and Plastic*; Hawkins, Potter, and Race, *Plastic Water*; Hawkins, "Performativity of Food Packaging"; Hodges, "Medical Garbage"; Klocker, Mbenna, and Gibson, "From Troublesome Materials to Fluid Technologies"; M. Liboiron, "Redefining Pollution and Action"; M. Liboiron, "Not All Marine Fish Eat Plastics"; Meikle, *American Plastic*; Pathak and Nichter, "Anthropology of Plastics"; Roberts, "Reflections of an Unrepentant Plastiphobe"; Huang, "Ecologies of Entanglement"; Helmreich, "Hokusai's Great Wave"; Gabrys, Hawkins, and Michael, *Accumulation*; Westermann, "When Consumer Citizens Spoke Up"; Wagner-Lawlor, "Poor Theory and the Art of Plastic Pollution in Nigeria"; and Stanes and Gibson, "Materials That Linger."

110 This section is based on a Twitter essay: M. Liboiron, "Good Question . . ."

111 Curry, *Digital Places*, 48.

relations are not properties of things so much as what make things. This text is from this place, and that means it will not always travel well, generalize well, make sense elsewhere (more on this in chapter 3). That's fine.

The province of Newfoundland and Labrador, and particularly the island of Newfoundland, was, and in many ways still is, a British colony that was stocked with Irish migrants to work as fish harvesters. The settler population is what is called "genetically isolated" or a "founder population," a rare condition that means that 98 percent of the settler population is genetically related.[112] Experientially, this means that the local accent is archaic Irish. Work holidays are Irish.[113] The food is Irish with a twist of cod. When the province joined Canada in 1949, the confederation document noted that there were no Indigenous people here and that, therefore, the Indian Act did not apply to the province.[114] This party line persists today despite the fact that the Bureau of Statistics recorded Inuit, Innu, and Mi'kmaq populations both before and after confederation. They were out and about buying bread, catching fish, going to school—but officially not existing.[115] So when I say Newfoundland and Labrador is a colony, I mean that it is characterized by a unique combination of remoteness, infrastructural sparseness, Indigenous erasure,[116] and settler homogeneity that shapes everyday lived experience, politics, and intellectual production.

Also in Newfoundland and Labrador: the Land is loud here, and settlers, Indigenous people (local and come-from-away), and others tend to notice their Land relations. On the west coast, 80 percent of the province's population eats local cod at least once a week,[117] and that percentage increases and the species diversify as you move north into Labrador.[118] When the cod fishery collapsed in 1992 after the introduction of Scientific fisheries management, it suddenly

112 Rahman et al., "Newfoundland Population."

113 St. Patrick's Day is a work holiday for government and university staff.

114 Hanrahan, "Lasting Breach." For the lasting repercussions of this on Indigenous nationhood, particularly for the Qualipu Mi'kmaq, see *The Country*, directed by Phyllis Ellis (Newfoundland, 2018).

115 Indigenous erasure isn't a new trick, nor is it unique to Newfoundland and Labrador. See Hall, "Strategies of Erasure"; Bang et al., "Muskrat Theories"; Barman, "Erasing Indigenous Indigeneity in Vancouver."

116 To expand on this idea, erasure doesn't end with recognition. For discussions of how settler-based modes of recognition can continue to erase Indigenous sovereignty and knowledge, see Coulthard, *Red Skin, White Masks*; and Anonymous Indigenous Authors, "Indigenization Is Indigenous."

117 Lowitt, "Examining Fisheries Contributions."

118 Durkalec, Sheldon, and Bell, "Lake Melville."

and acutely transformed the province.[119] The decline in the caribou population and resulting hunting ban in 2013 have likewise transformed Land and nation-to-nation relations in Labrador.[120] When I write about plastics and science, it is more than a case study: I'm talking about my food, other lab members' food (and often their families' histories and livelihoods), and the food, relatives, and heritage of Indigenous, settler, and other people in the province. I am beholden to all of them—these are my specific obligations as a scientist who works on plastics in wild food webs in Newfoundland and Labrador.

I can't talk about Land in Newfoundland and Labrador if I don't talk about the weather. Weather isn't small talk, as I learned when I first moved here and was trapped in my office when the snow outside reached up to my chest, or when I had to crawl home along the sidewalk in high winds so I wasn't blown into the road, or when ice pellets flying in 100 kilometre-per-hour winds made my face bleed, or that day no one came to work because it was sunny. The cabbies all talk weather and oil prices. They are what shape life here.

These Land relations keep me, and many others here, humble. Humility and modesty are different. Modesty means you don't talk about your accomplishments so that you don't elevate yourself over others. Humility means that you are connected to others, and it is the recognition that you cannot do anything without these many others, from the people watching your dogs, your kids, and your students so you can go to conferences, to the people who ensure that your water pipes and garbage cans and Internet work as intended. Cod, wind, snow, caribou—and plastics—are part of the others that connect people to one another and to Land here in Newfoundland and Labrador.[121]

These specific connections do not travel effortlessly to other places with other relations. This is one of the difficult parts of writing a book that travels more promiscuously[122] than the relations the book comes out of. You can read this tension, for example, in my discussion of the diversity of colonialisms, even

119 Bavington, *Managed Annihilation*.

120 Labrador Research Forum participants, "Caribou and Moose."

121 Sengers, "What I Learned on Change Islands"; Brynjarsdóttir and Sengers, "Ubicomp from the Edge of the North Atlantic."

122 *Promiscuous* is not my term for how written texts circulate willy-nilly. It's Plato's (unmarked). He thought that the written word could wander around and speak to whomever, regardless of whom the words were meant for, and this presented a real danger for love notes and other audience-based ethics. His text is performative of that fact, as he tries to get into the toga of a young man whose lover wrote him a love note that seems to have gotten into the wrong hands. Plato, *Plato's Phaedrus*.

as I often address colonialism as if it is fairly monolithic in most other parts of the book. The same holds for why I insistently differentiate between anticolonialism and decolonization—these insights and treatments come from stakes and contexts in Newfoundland and Labrador specifically and Canada more generally.[123] So, I ask of you, Reader, how do we write and read together with humility, keeping the specificity of relations in mind? How do we recognize that our writing and reading come out of different places, connections, obligations, and even different worldviews, and still write and read together?[124]

I was at an academic meeting when a settler researcher asked me and the Inuk next to me what we thought of Shawn Wilson's (Cree) *Research Is Ceremony*.[125] She patiently waited for our replies before telling us that she really couldn't see herself using it, that it was impractical for her kind of research. She said it wasn't for her. My initial response was that no research is exempt from the obligation of good relations, which is one way to understand what Wilson means by ceremony in research. But then it occurred to me that she was probably right. It wasn't for her. *Research Is Ceremony* is very Cree, by my reading. The relations discussed in it are rooted in Cree law, based on the "expectations

123 The Canadian spellings in the text are a reminder that these words come from somewhere.

124 One of the best methodological frameworks I've seen for reading with humility is Joe Dumit's (unmarked) "How I Read." I believe he wrote it in response to the dude-core practice of tearing texts apart as a dominant form of critical academic reading, particularly in graduate school. He outlines a variety of alternative ways to approach a text. I return to this work regularly to help remind me of the various ethics, aims, and collaborations possible in reading. Thank you, Joe Dumit, for your generosity of thought and social relations, and particularly how those things come together in your academic work. After hearing your talk "Elementary Relations: Bromine in Self, Society and World" in Barcelona, I was so inspired to write about relations that I left the conference early, booked myself into a hotel, and started writing this book on every paper surface I could find (coasters and napkins from the hotel feature prominently in the first draft of this text). Thank you and your co-panelists—Michelle Murphy, Dimitris Papadopoulos, Cori Hayden, and Stefan Helmreich—for that talk. Dumit et al., "Elements Thinking T122.1"; Dumit, "How I Read."

Drawing on Dumit's work, I've written about ethics and relationality in reading: M. Liboiron, "Exchanging."

125 S. Wilson, *Research Is Ceremony*. *Research Is Ceremony* is a foundational English-language text on academy-based Indigenous and decolonizing research methods. Thank you, Shawn Wilson, for being one of the early pathfinders for what a research text can look like if its format follows, as best as it is able, Cree law. To write a book as a letter to your family, writing in a way that makes extractive reading difficult and filling it with stories that are themselves analysis, is a gift in academic innovation. Also, I just found out Alex is your sister! So cool! Smith, *Decolonizing Methodologies*.

and obligations about proper conduct" that come from a particular place.[126] *Research Is Ceremony* is written as a letter to Wilson's sons. If relations are specific, then the methods simply will not work as well for anyone who is not Wilson's son. They might work a little, or even a lot, but relations do not universalize. To assume otherwise is not practicing humility with specificity. I'm pretty sure that's not what the settler researcher meant, but it was instructive nonetheless.

Like *Research Is Ceremony*, *Pollution Is Colonialism* is not written for or to everyone in the same way, or even at all. One of my primary struggles in writing this text is how it obliges me to different worlds and readers simultaneously. I am a scientist well seated in the domain of dominant science, even as I arrived via an academic trajectory in fine arts and media studies. I am also an STS-er, an anticolonial activist, and a scooped[127] and slowly reconnecting Métis/Michif.[128]

This text has been crafted and reviewed from similarly incommensurate standpoints. It has gone through academic peer review, first with brilliant friends and then with generous anonymous reviewers.[129] It has gone through

126 Borrows, *Canada's Indigenous Constitution*.

127 Kimmelman, "No Quiet Place."

128 A primer on terms! Because terminology stems from settler government legislation as well as the self-determination of Indigenous groups, terms are always shifting. Different terms are used at different historical moments, in different places, and by different groups and governments. At this time, *Indigenous* is a term used by the United Nations to mean all first peoples around the world. It's also a common term in academia, though often not in communities. It is not a perfect term, but it is the term that applies to the broadest number of peoples and is legible to the broadest number of researchers at this moment. It's the term I use in this text for this reason. In the next book, that might change.

Aboriginal is a term that comes from Canada's 1982 Constitution (section 35), and it refers to all First Nation, Métis, and Inuit groups in Canada. This does not mean, however, that it is embraced by all groups.

First Nations refers only to groups included within Canada's Indian Act (1876) and does not include Métis or Inuit.

For an overview of this terminology in Canada, see Vowel, *Indigenous Writes*.

For more on the complexities of *Métis* as a term whose racial formations we are constantly fighting, particularly in Atlantic Canada where the term has been racialized and appropriated in different ways, see Andersen, *Métis*. For these reasons, here in Atlantic Canada I use the term *Michif*. Thank you, Chris Andersen (Métis/Michif), for your book and for your comradery, generosity, and jokes that kept me planted in a shared space even when we're geographically far apart. Maarsi.

129 Dear anonymous reviewers: Thank you for your time, your labour, your generosity, your work to make this book work better. Duke University Press helped ensure that different reviewers came from different readerships, and your insights helped me see how different

Elder review to ensure the text did not stray from good relations, both in terms of speaking truth to shared Indigenous laws, values, and knowledge, and not overstepping what could be shared. Scientists and anticolonialists, Elders and peer reviewers do not necessarily agree on what is true and right and good. These positions are incommensurate: they do not share a measure of value. This text is beholden to all of them, to its readers, to its place, and thus to multiple incommensurabilities.[130] In "Decolonization Is Not a Metaphor," Tuck and Yang write that "an ethic of incommensurability . . . recognizes what is distinct"[131] and what cannot be joined or conflated. It "brings these areas into conversation, without papering over the differences, but also without maintaining false dichotomies."[132] In this book, there are moments when different kinds of readers are called out, called in, and called down to the footnotes. There are moments that might appear contradictory, at odds, or mutually exclusive because they are.

As collaborators Alison Jones and Kuni Jenkins (white/settler/Pakeha and Maori/Ngati Porou) have written:

> Research in any colonized setting is a struggle between interests, and between ways of knowing and ways of resisting, and we attempt to create a research and writing relationship based on that tension, not on its era-

---

audiences need different things in a text and how I might balance those differences and inclusions. Thank you.

130 Science historian Thomas Kuhn (unmarked) talks about the "incommensurability of competing paradigms. In a sense that I am unable to explicate further, the proponents of competing paradigms practice their trades in different worlds. . . . Practicing in different worlds, the two groups of scientists see different things when they look from the same point in the same direction. Again, that is not to say that they can see anything they please. Both are looking at the world, and what they look at has not changed. But in some areas they see different things, and they see them in different relations one to the other. That is why a law that cannot even be demonstrated to one group of scientists may occasionally seem intuitively obvious to another. Equally, it is why, before they can hope to communicate fully, one group or the other must experience the conversion that we have been calling a paradigm shift. Just because it is a transition between incommensurables, the transition between competing paradigms cannot be made a step at a time, forced by logic and neutral experience. Like the gestalt switch, it must occur all at once (though not necessarily in an instant) or not at all. How, then, are scientists brought to make this transposition? Part of the answer is that they are very often not." This book is written from different worlds, if you will, and has these same issues. Kuhn, *Structure of Scientific Revolutions*, 150.

131 Tuck and Yang, "Decolonization Is Not a Metaphor," 28.

132 Tuck and Yang, "Decolonization Is Not a Metaphor," 5.

sure. Indeed, we seek to extend the tension, and examine its possibilities. In doing this, we cautiously reject the usual suggestion that indigenous-coloniser/settler research relationship should be based in "mutual sharing," or "understanding," or even collaboration when understood in such terms. These injunctions can be understood as calling on certain postures of empathetic relating which aim at dissolving, softening or erasing the hyphen, seen as a barrier to cross-cultural engagement and collaboration.[133]

Often this ethic of incommensurability "limit[s] what we feel free to say, expand[s] our minds and constrict[s] our mouths . . . within the negotiated relations of whose story is being told, why, to whom, with what interpretation, and whose story is being shadowed, why, for whom, and with what consequence."[134]

For this reason, there are many things not said in this text. First, you'll notice the book is about colonial systems of science and pollution, not about the ways Indigenous peoples are disproportionately harmed by pollution. Following Audra Simpson (Mohawk), "I refused then, and still do now, to tell the internal story of their struggle. But I consent to telling the story of their constraint."[135] Along with Eve Tuck, I refuse to reproduce "damage-centered research . . . that operates, even benevolently, from a theory of change that establishes harm or injury in order to achieve reparation" and instead work to put "the context of racism and colonization" at the centre of pollution research.[136] I follow the call to focus on colonialism, rather than its effects, sounded by Aileen Moreton-Robinson and others when they call for research to move Indigenous studies "beyond identity concerns to develop and expand its mode of inquiry to a range of intellectual projects that 'structure inquiry around the logics of race, colonialism, capitalism, gender and sexuality.'"[137] I'm following many people who have elevated refusal into a practice of affirmation, repair, and resurgence, looking upstream to see structures of violence rather than effects and harm.

These methodological strategies and negotiations are usually written about as methods for researchers and writers, but I would ask that *readers* take them up as well.[138] If at some point, as you read, you think "this isn't for me, I can't take this up," you may be right, but that response does not foreclose the invitation

133 Jones and Jenkins, "Rethinking Collaboration," 475.
134 Fine, "Working the Hyphens," 72.
135 A. Simpson, "Consent's Revenge," 328.
136 Tuck, "Suspending Damage," 413, 415.
137 Moreton-Robinson, *White Possessive*, xvii.
138 M. Liboiron, "Exchanging."

to keep reading. It is an occasion to ask what is happening between[139] yourself and the text. Reading ethically can mean refusing to read as a form of extraction, though academia has trained us to do so. Tuck has written:

> To watch the white settlers sift through our work as they ask, "isn't there more for me here? Isn't there more for me to get out of this?" . . .
>
> Isn't there something less theoretical? Something more theoretical? Something more practical? Something less radical? More possible?
>
> Can't you make something that imagines it clearly enough for me to see it? For me to just plunk it into my own imagination?
>
> Can't you do more work for me? because I have given this five whole minutes of thought and I don't see the future like you. . . .
>
> I'll just keep sifting through all of this work that was never meant for me, sorting it by what is useful to me and what is discardable. . . .
>
> I forgot that people read extractively, for discovery[.]
>
> I forgot that all these years of relation between settler and Indigenous people set up settlers to be terrible readers of Indigenous work.[140]

The first time I read this thread it shocked me into reflexivity because, while I try to stay in good relations, I often—usually—read extractively, looking for bits I can use. I had been reading in a Resource relation (see chapter 1) that is unidirectional, assessing texts solely for my own goals and not approaching them as bodies of work, events, gifts, teachers, letters, or any number of other ways that would make unidirectional, extractive relations seem rude and out of place.

As a writer, I have tried to write less extractively by citing at length, footnoting my relations to texts, leaving things out, and spending considerable time on certain concepts to balance obligations to different audiences and knowledge systems. I've also tried to support readers in reading less extractively by addressing the reader explicitly, using jokes to make space for difficult concepts, being clear that this is a text written out of the province of Newfoundland and Labrador, and signaling how not all ideas travel effortlessly and easily root in other places. You don't need a lab like CLEAR to attune everyday intellectual practices to anticolonialism. Writing and reading are relations. We have already started.

139 Fine, "Working the Hyphens."
140 Tuck, "To Watch the White Settlers . . ."

A Road Map

This is a methodological text, where methodology is understood as a way of being in the world. An ethic, if you like that word better. There are colonial ways to be in the world, whether intentionally or otherwise, and there are less colonial and anticolonial ways to be in the world. This includes science. Throughout this book, I redefine pollution as central to, rather than a by-product of, colonialism, and I think about the role of science in achieving both colonialism and anticolonialism. I use plastics and their status as a pollutant to investigate and then refute those colonial relations. Often, I'll turn to CLEAR as the lens and framework to denaturalize colonial scientific practices and concepts of land, Nature, and Resource, while also giving examples of anticolonial science and methodologies that produce diverse futures. As such, this text is less about claims and more about models. I hope the text is useful to you. But not in a creepy, Resource-y way.

The first chapter, "Land, Nature, Resource, Property," outlines the historical and conceptual groundwork for the invention of modern environmental pollution as a colonial achievement. It discusses Indigenous concepts of Land and how these ideas get flattened into Nature through colonial relations based in separation, universalism, and the scientifically proven resilience of the natural world. Building on these concepts, I theorize Resource relations, by which I mean the morality of maximum use of Resources, dispossession, and property as a way to control both time and space to secure settler and colonial futures. This mode of Resource relations is a hallmark of colonialism. Two story lines animate this discussion. The first is the story of Streeter and Phelps's pioneering[141] work on assimilative capacity that defined the moment of pollution as that when bodies of water could no longer assimilate pollution. Everything else was mere contamination. A second story interrupts the first with short vignettes from CLEAR, as lab members grapple with legacies of colonial science as well as events, practices, relations, and landscapes that refuse logics of colonial relations.

The second chapter, "Scale, Harm, Violence, Land," builds out plastics as more than a monolithic pollutant that must be banned or eradicated, not as a theoretical exercise, but for the purpose of working with plastics in science and activism. I theorize scale as a way to understand specific relationalities, differ-

141 "OMG. Why do you flag *pioneer* every time? We get it. It's a dirty word." I flag it because dirty words are not to be left unattended. That's how they get laundered and normalized. Bad *pioneer*.

ences between harm and violence, and recourses to purity in environmental activism and dominant science. I recount how settler endocrinologists, conservationists, and toxicologists come to understand plastics and their chemicals in complex ways that open up dominant science as a practice already rife with examples of and impulses to anticolonial work, troubling the division of Nature from humans, the autonomy and discreteness of both matter and agency, and universalism. The chapter closes with examples from Indigenous thinking about plastics as Land to extend anticolonial framings of plastics' diverse L/land relations (please fully read that part of the chapter if you just started salivating at the phrase "plastics as Land").

The third and final chapter, "An Anticolonial Pollution Science," lays out the *how* of CLEAR's anticolonial science via our methods. I use the examples of CLEAR's unique practices of peer review and sampling to return to concepts of specificity and obligation. I introduce the framework of compromise to describe some ways to ethically maneuver the uneven power relations of dominant and anticolonial science. It ends with final thoughts on how to stay true to critiques of universalism while also generalizing the lessons of the text into what I imagine to be the Reader's own work—How do place-based, nonuniversal methods travel? How do we take messages with us without being extractive or Resource-oriented? How do they become useful and good in other places, for other people, like you? I look forward to the stories you tell[142] when you stand on CLEAR's shoulders. You might think of this final chapter as dessert. Sometimes I eat dessert first. But the book as a whole ensures that the last chapter is not just delicious but not-very-nutritious sugar. Together, the chapters build up the nuances, stakes, and methodological legacies that ground CLEAR's work.

This is a book about work. Really hard work. I'm always glad when people raise a fist against the injustices of systems, including pollution and its sciences. But I'd much prefer people pick up a shovel—or a microscope—with the other hand and get to work. *Pollution Is Colonialism* is designed to show how scientists and others are already working in an anticolonial way. We always already are in L/land relations, and they come out in our methods. Time to start.

142 And cite! One of the issues we face in CLEAR regularly that I'll bring up again in chapter 3 is being thanked for our work and how it helps others in their own research, but our intellectual production is not cited. Please follow basic academic manners and cite methods I am sharing, which have been proposed, tested, tweaked, validated, and laid out here after peer review. Thank you.

1 · Land, Nature, Resource, Property

Permission to Pollute

According to Canadian federal regulations, 0.010 milligrams (mg) of arsenic per litre (L) of drinking water is acceptable, but 0.011 mg/L is too much.[1] The maximum acceptable concentration for lead in tap water is 0.005 mg/L.[2] Under the permission-to-pollute system, specific quantities of contaminants are allowed legally in bodies of water, human bodies, air, food, and environments. This way of governing pollution is relatively new, but it is premised on an old colonial system of land relations where the land is a Resource.[3]

A core scientific achievement in the permission-to-pollute system was the articulation of *assimilative capacity*—the theory that environments can handle a specific amount of contaminant before harm occurs. Today, assimilative capacity is a term of art in both environmental science and state regulation. It refers to "the amount of waste material that may be discharged into a receiving water without causing deleterious ecological effects."[4] Measures of assimilative capacity compare the rate of metabolism with the rate of pollution, with assimilative

1 Health Canada, "Guidelines for Canadian Drinking Water Quality."

2 Health Canada, "Lead in Drinking Water."

3 This argument was first articulated in a pamphlet titled "Pollution Is Colonialism" created by my lab and another group I'm part of, and it relies heavily on my and Michelle Murphy's work. See CLEAR and EDAction, "Pollution Is Colonialism"; and Shadaan and Murphy, "EDCs as Industrial and Colonial Structures."

4 Novotny and Krenkel, "Waste Assimilative Capacity Model for a Shallow, Turbulent Stream," 604.

capacity marking the place where the two are equal. This is the threshold of harm. The threshold theory of pollution differentiates between *contamination*, as the mere presence of a pollutant, and *pollution*, as the manifestation of (scientifically!) demonstrable harm by pollutants when metabolism is overwhelmed.

Assimilative capacity is based on land relations that strip away the complexities of Land[5]—including relations to fish, spirits, humans, water, and other entities—in favour of elements relevant to settler and colonial goals for using the water as a sink, a site of storage for waste.[6] As la paperson (diaspora settler of colour)[7] has written, "Primitive accumulation involves not only the gathering of 'natural' resources as assets but also the externalizing of the 'cost' of the accumulation in the form of contaminated water, disease, and other traumas to the 'natural,' nonpropertied, that is, 'Indigenous,' world. To be subject to anti-Indian technologies does not require you to be an Indigenous person."[8] Sinks are one such "anti-Indian," colonial technology.

Assimilation theory transforms bodies of water and other environments into a Resource for waste disposal. As I will argue in this chapter and throughout the book, this kind of environmental science is premised on access to Land for settler and colonial goals (in this case, waste disposal). Through dominant science[9] and other methods, these land relations come to seem true, good, and natural.

This and subsequent chapters are structured so that theoretical and historical arguments are interspersed with memories and stories from CLEAR lab members. Some of these stories are mine, but most aren't. They are designed to clarify and expand on ideas but also to show that science, colonial or anticolonial, is not monolithic and points of friction and opportunities for doing science otherwise present themselves regularly. All stories are shared with permission.

5 See footnote 19 in the introduction for why *land* is sometimes capitalized and sometimes not.

6 A sink is a land-based place to store waste. In the words of historian Joel Tarr (unmarked), "Much of the history of industrial waste disposal, as well as the disposal of wastes from other sources such as an urban population, involves the search for a 'sink' in which wastes could be disposed of in the cheapest and most convenient manner possible." Tarr, "Searching for a 'Sink,'" 9.

7 See footnote 10 in the introduction for why some authors are (unmarked) and others are (diaspora settlers of colour).

8 paperson, *A Third University Is Possible*, 11. For readers who might be new to how Indigenous theory extends to non-Indigenous people, and how colonialism is a set of land relations that can travel to places that may or may not have Indigenous peoples, the last sentence in this quote is for you.

9 See footnote 77 in the introduction on why I use the term *dominant science* instead of *Western science*.

Not All Pollution Is Colonialism

The land relations behind the definition of pollution as the moment after assimilative capacity is surpassed have become so naturalized that other ways of discarding, based in other worldviews, have been obfuscated and even eliminated. When most people refer to waste and pollution today, they are referring to a set of relations that uses Land as a sink for a relatively new form of waste characterized by unprecedented tonnage, toxicity, and heterogeneity,[10] created within industrial political economies premised on growth and profit. But not all forms of pollution and waste are colonialism.

The settler woman is telling us (a group of Indigenous delegates) about how she admires that Indigenous people use the whole animal. I immediately think of the seal hunting[11] trip that CLEAR's Inuk community coordinator just returned from to gather food and scientific samples. On her first day back to the lab I asked if they got sealskins. She said that beautiful as they are, the skins are too hard to remove and prepare these days, so they took the meat, gastrointestinal tracts, jaw bones, and biopsies and let the rest of the seal slide back into the water. They certainly didn't "use" the whole animal. This didn't bother either of us: the seal will be used by other animals under the ice. In fact, when we're done researching the seal guts, we'll return them to the Land to feed our relatives as well. I decide not to tell the woman any of this. After all, she's not exactly wrong, and I have no energy for nuance today.

Other ways of discarding have been obfuscated by the dominance of modern, Eurocentric meanings.[12] From a scientific perspective, discarding seals and

10 This characterization of modern waste is from MacBride, *Recycling Reconsidered*.

11 Public service announcement from Newfoundland and Labrador and Inuit Nunangat (Inuit homelands) generally: seal hunting is not only legal and regulated in Canada, but a respected and respectful part of Inuit food webs, relations, and lifeways. Just as modern waste is characterized by massive tonnage, toxicity, and heterogeneity and is completely different than other forms of waste, so, too, does sustenance-based, traditional seal hunting differ from industrial seal hunting. If you're struggling with these ideas, I recommend Arnaquq-Baril, *Angry Inuk*. For me, seal hunting was the first blatant case where I fully understood that all feminisms are not also anticolonial. Awkward in a feminist *and* anticolonial lab.

12 One of the academic fields I work in is discard studies, the social science of waste and wasting (see discardstudies.com). I don't think I've come across a definition of waste in an academic space that doesn't either centre the human or conflate all nature and animals (and occasionally Indigenous people) as one kind of thing that wastes "naturally." If you know of any research that offers more nuance, please let me know.

discarding plastic packaging are materially different and have fundamentally different effects in ecological systems.[13] From an anticolonial perspective, the land relations that result in their discard are also different, based on fundamentally different L/land relations. Let's not conflate those differences.

Modern Environmental Pollution Is Colonialism

When I began this work, I wondered why modern environmental pollution was so easily economized. What allowed such complete capture of environmental regulation by industry to exist from the earliest moments of the twentieth century? It is not enough to say that industry and government's ability to pollute is a logical strategy to achieve the twinned pursuits of growth and capital. It is insufficient to say that Nature was understood by scientists to be a silver bullet to solve waste problems. What allowed these things to make sense in the first place? Why was not only the ability, but also the imperative, to pollute on the table at all? Under what conditions does managing, rather than eliminating, environmental pollution make sense?

That would be colonialism.

It's time to go deeper. Let's start with the basics: Land, Nature, and Resource.

Land

Defining Land by typing it out onto a page is like defining your favourite aunt as your mother's sister. True, yes, but your favourite aunt is more than that—she is the host of giant spaghetti meals and countless hours at the kitchen table teaching you how to draw horses. She is the one to tell you not to go with that man because he's no good.[14] She is the promise that someone will take care of you if something happens to your parents. So, too, with Land. If you'd like to learn more about Land, I recommend reading botanist Robin Wall Kimmer-

13 This is not just true of seals and plastics, but also of plastics and plastics. In 2015, I wrote, "The difference between PET plastics used in soda pop bottles and PVC plastic used for water pipes matters because the materials fragment, travel, and influence bodies differently. It matters whether that PET or PVC is in water, in a cod stomach, or on a store shelf because it will cause harm differently, and cause different types of harm, in each case." I have learned a lot since then, and the analysis could have used critiques of agency laid out by Vanessa Watts (Anishinaabe and Haudenosaunee). But the lesson about specificity and materiality in relations still holds and has only gotten stronger over time. M. Liboiron, "Redefining Pollution and Action," 5. See also Watts, "Indigenous Place-Thought."

14 For the academic equivalent, the "academic auntie," see E. Lee, "I'm Concerned for Your Academic Career If You Talk about This Publicly."

er's (Potawatomi) *Braiding Sweetgrass*, which beautifully narrates the ethos of Land though rich and interwoven narratives of knowledge, action, accountability, and beings.[15]

Defining Land makes it sound like a noun. But Land is a verb: "A bay is a noun only if water is *dead*."[16] Collaborators Eve Tuck (Unangax) and Marcia McKenzie (settler) write that Land "is both a notion and an action."[17] Land never settles. It is about relations between the material aspects some people might think of as landscapes—water, soil, air, plants, stars—and histories, spirits, events, kinships, accountabilities, and other people that aren't human.[18] These relations are happening all at once rather than being parceled into individual paired units, like plant to soil, mother to daughter. We have some plant mother soil plant mother going on.

Robin Wall Kimmerer writes that Land is "everything: identity, the connection to our ancestors, the home of our nonhuman kinfolk, our pharmacy, our li-

15 Thank you, Robin Wall Kimmerer. I first encountered your name as a signatory of the Indigenous Science Statement for the March for Science (http://www.esf.edu/indigenous-science-letter/). I was looking for other Indigenous scientists to learn with from afar. That's when I discovered the immense gift of your books, including *Braiding Sweetgrass*, which wove together law, Land, and science. It set the bar high, beautifully. Maarsi.

16 Kimmerer, *Braiding Sweetgrass*, 55; emphasis in original. Taking this further, la paperson writes, "The subjugation of land and nonhuman life to deathlike states *in order* to support 'human' life is a 'biopolitics' well beyond the Foucauldian conception of biopolitical as governmentality or the neoliberal disciplining of modern, bourgeois, 'human' subject [*sic*]." He adds, "The exercises of supremacist sovereign power over life and death are most chillingly undisguised when we consider the ways the life worlds of land, air, water, plants and animals, and Indigenous peoples are reconfigured into natural resources, chattel, and waste: statuses whose capitalist 'value' does not depend on whether they are living or dead but only on their fungibility and disposability. For example, in modern animal industrial processes, the carcass is valued just as much as, if not more than, the breathing animal." paperson, *A Third University Is Possible*, 5, 14–15. I never thought that reading a book about universities would be so rich with anticolonial environmental thought. Thank you, la paperson, for your words, your work, your collaborations, and your commitments to alliances against colonialism in its varied forms.

17 Tuck and McKenzie, *Place in Research*, 57.

18 Like many others, I struggle with the designation of *nonhuman* to mean everyone who isn't human since the term recentres the human at the moment you're trying not to. It's like calling all people who are not men, nonmen. The alternative *more-than-human* also leaves humans in the middle, though I appreciate its commitments. In everyday speech I tend to say "people," but then humans often think I'm talking about them again. I will use various terms in this text in an attempt to make as much sense as I can in any given moment. For more, see Chagani, "Can the Postcolonial Animal Speak?"

brary, the source of all that sustains us. Our lands were where our responsibility to the world [is] enacted."[19] In *Native Science*, science philosopher Gregory Cajete (Tewa, Santa Clara Pueblo) talks about relations with/in Land as "ensoulment," "a kind of a map of the soul" where the soul of the Land and of people are the same thing.[20] Enrique Salmón (Rarámur, Tarahumara) explains: "When [Indigenous] people speak of the land, the religious and romantic overtones so prevalent in Western environmental conversation are absent. To us, the land exists in the same manner as do our families, chickens, the river, and the sky. No hierarchy of privilege places one above or below another. Everything is woven into a managed, interconnected tapestry. Within this web, there are particular ways that living things relate to one another."[21]

This is CLEAR member Charles Mather's (settler) memory about learning a little about complex Land relations:

> *During the recreational food fishery, when people in Newfoundland can fish legally, we go to the wharfs where people are filleting their fish and ask for the fish guts. Sometimes commercial fishers are also at these docks, and we ask them as well. We get hundreds of guts this way every season. This approach to sample collection aligns well with our commitment to accountability and to good relations. Legally, cod must be gutted on the wharf and cannot be processed at sea. Fishers, both commercial and recreational, typically discard the guts into the sea around the wharf, keeping the tasty fillets, cod cheeks, and britches. So our samples come from something that has to be processed on the wharf, is not normally used by humans, and would have been thrown away. This is good.*
>
> *But in our second year of sample collection on the wharfs around St. John's, someone else wanted cod guts. She wasn't interested in the guts for scientific purposes. She didn't want to know how much plastic the cod had ingested. Instead, she wanted to use the cod carcasses to make a soup or a broth. We thought the cod carcasses and guts were waste, but clearly that is not the case for everyone. We were surprised and taken aback. What had seemed such an ethically uncomplicated way of collecting samples had suddenly become deeply complicated. How could we take food away from someone in order to generate data in our lab? That didn't align well with our commitment to good relations.*

19 Kimmerer, *Braiding Sweetgrass*, 17.
20 Cajete, *Native Science*, 133.
21 Salmón, *Eating the Landscape*, 27.

This encounter has fundamentally changed the way we relate to people and fish on the wharf. We no longer work from the premise that the carcasses and guts are waste, to which we have exclusive and uncomplicated access. We approach sample collection more cautiously and with the sensibility that we cannot know with certainty all of the ways in which fish will be used. And we recall this experience with existing and new lab members to illustrate our commitment to humility in our scientific research practices.[22]

See how tricky colonialism is?[23] Settler access to Land gets in at every turn: *We no longer work from the premise that the carcasses and guts are waste, to which we have exclusive and uncomplicated access.* Colonialism lurks in assumptions and premises, even when we think we're doing good.

The Land lesson I want to focus on here is about specificity. Unlike land, Land is fundamentally relational and is *specific* to these relations: "Every cultural group established this relationship to [their] place over time. Whether that place is in a desert, a mountain valley, or along a seashore, it is in the context of natural community, and through that understanding they established an educational process that was practical, ultimately ecological, and spiritual. In this way they sought and found their life."[24] This refers not only to the differences between St. John's and Toronto, but also to the relational differences between being a researcher and a fish harvester in St. John's, even if the researcher and fish harvester are the same person.[25] This is why Land is capitalized—it is the

22 This story was first written down in 2017 but occurred earlier, likely in 2015. Charlie, thank you for everything you have done for CLEAR and for me. Your presence as a full professor working within a young assistant professor's lab is not only an expression of CLEAR's commitment to different forms of knowledge and experience, but you have provided an invaluable sounding board, acted as a co-mentor, and been a source of enthusiasm for the lab during the difficult, complicated, and sometimes demoralizing (but also exciting, invigorating, and beautiful!) process of directing a feminist and anticolonial lab. Thank you! Readers, get this: Full professor Dr. Mather (and full professor Dr. Power) are not co-leads of CLEAR. They are members. They outrank me and work under my direction simultaneously, collaboratively, and this is good.

23 BTW, this doesn't make colonialism a trickster. If tricksters aren't yours, leave them alone. They will kick your ass.

24 Cajete, *Native Science*, 133.

25 This is a hard lesson to teach Indigenous people who become graduate students and thus researchers: as a researcher, you have fundamentally different obligations, relations, and legacies to account for than when you are just an Indigenous person fishing. You are part of the academy, a colonial project, and you are reproducing those relations even as you seek to change them. For example, research usually aims to generalize knowledge, even though

shorthand for all these relations as a proper name that is specific and unique, not universal and common.

Science Happens on Land

Science always happens within land relations, and those relations are always specific to that place, even if you don't believe in Land. The theory of assimilative capacity makes claims to universality, but the scientists who developed it, H. W. Streeter and Earle B. Phelps, were working in a specific place and they needed particular Land relations for their universal theory to work. They were looking at the Ohio River in the Ohio River Valley, which was not only saturated with municipal organic waste, but also a site for large-scale disposal of coal-tar waste. The interaction between the coal-tar wastes and the chlorine used to disinfect municipal water supplies made the water taste and smell repugnant. The result was that people were choosing to drink more palatable, untreated water over the disgusting-tasting treated water, even after the germ theory of disease was widely accepted. The resulting disease outbreaks came not from pollution so much as locals' choice to opt for drinking (palatable) polluted water despite the availability of (disgusting) potable water.[26] At any rate, the federal government saw the Ohio case as a unique crisis that chlorine treatment was no longer addressing and sent scientists to the river to see what could be done.

Phelps turned to the Ohio both because of his interests in public health and because the river's sluggish, wide, densely polluted waters offered laboratory-like conditions to prove his theory of assimilative capacity. An earlier attempt to

---

in many cultures only Elders are supposed to make those sorts of bold statements. Your family and community members are now research subjects, collaborators, and beneficiaries of your work (both harm and benefit). Those are new relations. Some are incommensurate with existing relations. It is damn tricky to do research as an Indigenous person in your own communities. Thank you, Eve Tuck, for bringing these ideas up at the Labrador Research Forum in 2019 and in ongoing conversations, and to Ashlee Cunsolo for working on them with me for our students. See Tuck, "Research on Our Own Terms."

26 Historian Joel Tarr (unmarked) describes how, "in the mid-1920s[,] the U.S. Public Health Service . . . identified 25 cities in the Ohio River Valley where the interaction of chlorine with phenol wastes made the water almost undrinkable," and he links these aesthetics to "a typhoid outbreak of eighteen cases and three deaths in 1925 in Ironton, Ohio." Tarr, "Industrial Wastes and Public Health," 20. For more on the important role of aesthetics in pollution, despite scientific efforts to eradicate "subjective" aesthetics from understandings of pollution, see Christy Spackman's (unmarked) work: e.g., Spackman and Burlingame, "Sensory Politics."

prove assimilative capacity (what Phelps called self-purification at the time) in New York City's harbour had yielded promising results, but "there is evidently something wrong with our values at the region of Hell's Gate. . . . [A]n influence of the Harlem River shown here . . . was not properly taken into account in our computation."[27] Harlem will do that. While Phelps argued that his "general formula and hypotheses [of the ability of rivers to self-purify themselves of pollutants] are substantially accurate,"[28] he also knew that the noisy, not-very-laboratory space of New York Harbour was too tangled to emphatically prove his theory of self-purification. He needed the slow and straight Ohio River.

The concept of self-purification was first developed and studied in Britain, but it became a dominant scientific fact in the United States only after the work of Phelps, Streeter, and others. In 1860, British scientist Edward Frankland (unmarked), an expert responsible for keeping London's drinking water disease-free, noted, "There is no river in the United Kingdom long enough to effect the destruction of sewage by oxidation."[29] Outbreaks of typhoid at the time seemed to confirm this. I'd like to think it was the great lengths of American rivers compared to short and overpopulated British ones that allowed self-purification and the threshold theory of pollution to be further developed by Streeter and Phelps.

By articulating the Ohio River as a proper sink for pollution, Phelps and Streeter transformed it from a complex set of relations to one consisting of only a few relevant factors, what Jackie Price (Inuk) calls a "metaphysical flattening."[30] This is, as we shall see, a theme in colonial science's approach to Nature. The tension between universalism and emplacement—attempts to order and control Nature despite Land's relational specificity—offers me an opportunity to denaturalize the land relations of dominant science in favour of other ways of relating to the world, even (and especially) when dominant science seems to have a monopoly on describing land relations.

27 Phelps, "Chemical Measure of Stream Pollution," 533. Despite the disruption of data around Hell's Gate, Phelps held that the overall dataset still aligned with his theory of self-purification. But it wasn't quite enough to constitute a theatre of proof that made his claim so self-evident that the truth could speak for itself. The term *theatre of proof* comes from Latour, *Pasteurization of France*.

28 Phelps, "Chemical Measure of Stream Pollution," 533–34.

29 Rivers Pollution Commission, "Sixth Report," 138. For excellent research on the scientific struggles of water purification in the context of governance in Britain, see Hamlin, *Science of Impurity*.

30 Jackie Price, "But You're Inuk, Right?"

Nature

Small-*l* land is usually synonymous with Nature, in that both focus on only some aspects of relations, such as soil, air, water, animals, and plants, but not on human people, events, memories, spirits, or obligations.[31] Nature describes colonial relations with capital-*L* Land.[32] Whether Nature is understood as wild and heartless, the helpless victim of industrial assault, or the raw stuff of scientific enquiry, one of Nature's defining characteristics is that it is separate from humans, even if there is a closeness or affinity between them.

Separation

The naturalization of separation allows the scientific logic of variables to make sense—variables are ways to treat elements of an environment as discrete, autonomous actors.[33] For Streeter and Phelps, the variables that mattered to their theory included the flow rate, volume, temperature, time, and oxygen levels of the water, all of which were related but could nevertheless be separated from one another and independently measured.

For Streeter and Phelps, the key Natural phenomenon under study was self-purification, a phenomenon they believed could be graphed and used to predict

31 A "construction of nature" analytic flourished in the late 1990s and early 2000s, characterized by studies that examined how human-environment relationships became constructed via epistemology, cultural discourses, economic structures, and more. The when, how, and why of Nature's separation from (some) humans has been written about extensively, as, for example, in the following works: Merchant, *Reinventing Eden*; Schiebinger, *Nature's Body*; Jennifer Price, *Flight Maps*.

32 The terms *land*, *nature*, and *resource*, with or without capitals, are used in a variety of settings, including Indigenous ones. I am not the word police: people can use any of these terms to mean what they're trying to say. Many of us use the language of colonizers to mean things that exceed colonial worldviews (that's part of what colonization does!). If you're using these terms in your tribal council, traditional teachings (including Natural Law), writing, or everyday speech with other Indigenous people, keep on if it's working. I'm setting up these terms within this book so I can be clear about what I mean when I use them. Thank you, Kim TallBear, for conversations about words and their multiple meanings, and specifically about Elders who speak their traditional language and use the term *Creator* in their prayers—they might mean the Christian God, but they might also mean that-mystery-that-things-come-out-of-that-is-probably-related-to-stars-but-how-could-we-possibly-know-since-that-would-be-rude. Or similar.

33 This isn't to say that variables are "wrong" or fundamentally colonial things (they don't automatically grant access to Land for colonial goals), but to point to how worldviews allow some things to make sense and act as truth at the expense of other things.

when a river can no longer reintroduce enough oxygen to metabolize organic waste. What didn't matter to this universal phenomenon were things like smell, fish health,[34] water colour, or whether the river was nice to swim in, all of which had been used to define pollution in the past. The Streeter and Phelps equation implicitly argued that smells, fish, and swimmability, among other relations, were not essential characteristics of river pollution. Instead, oxygen levels and their variables became the focus of purification and thus of its opposite, pollution.

After empirically testing their variables, Streeter and Phelps arranged them in fixed and predictable relations—that is, they built a model. For each variable, they used measurements from the Ohio River to determine when the oxygen demand of metabolizing waste exceeded the reintroduction of oxygen into the stream (called re-aeration). Then they graphed the results.

There it is! You can see it in black and white![35] The failure of the river to purify itself is represented, over and over again, in the droop of the curve. The point at which each curve droops and flatlines for a moment is the point at

34 One source of pushback against Streeter and Phelps's equation concerned fish health. In 1933, fisheries scientist Carl L. Hubbs (unmarked) argued that fish health could not be totally captured by oxygen rates, and that the time of day, an essential characteristic that impacted fish health, was absent from Phelps's variables. But this critique, as well as others, did not reject the premise that a complete set of essential characteristics for pollution could be mathematically articulated. Instead, these critiques argued for increasing the number of variables and complexity of the pollution models while still maintaining that Nature and its pollutants could be understood and intervened in. Hubbs, "Sewage Treatment and Fish Life."

35 Here is the theatre of proof! Latour (unmarked) talks about the role of theatrical charisma in changing scientific paradigms when he tells the story of Louis Pasteur and his highly unlikely microbes: "Pasteur's genius was in what might be called the theater of the proof. Having captured the attention of others on the only place where he knew that he was the strongest, Pasteur invented such dramatized experiments that the spectators could see the phenomena he was describing in black and white. Nobody really knew what an epidemic was; to acquire such knowledge required a difficult statistical knowledge and long experience. But the differential death that struck a crowd of chickens in the laboratory was something that could be seen 'as in broad daylight.' Nobody knew what spontaneous generation was; it had given rise to a highly confusing debate. But an elegant, open, swan-necked bottle, whose contents had remained unalterable until the instant the neck was broken, was something spectacular and 'indisputable.'" Latour, *Pasteurization of France*, 85. Thomas Kuhn also talks about the central role of persuasion and beauty in shifting scientific paradigms: "Because [paradigms] differ about the institutional matrix within which political change is to be achieved and evaluated, because they acknowledge no supra-institutional framework for the adjudication of revolutionary difference, the parties to a revolutionary conflict must finally resort to the techniques of mass persuasion." Kuhn, *Structure of Scientific Revolutions*, 93. So, too, with pollution.

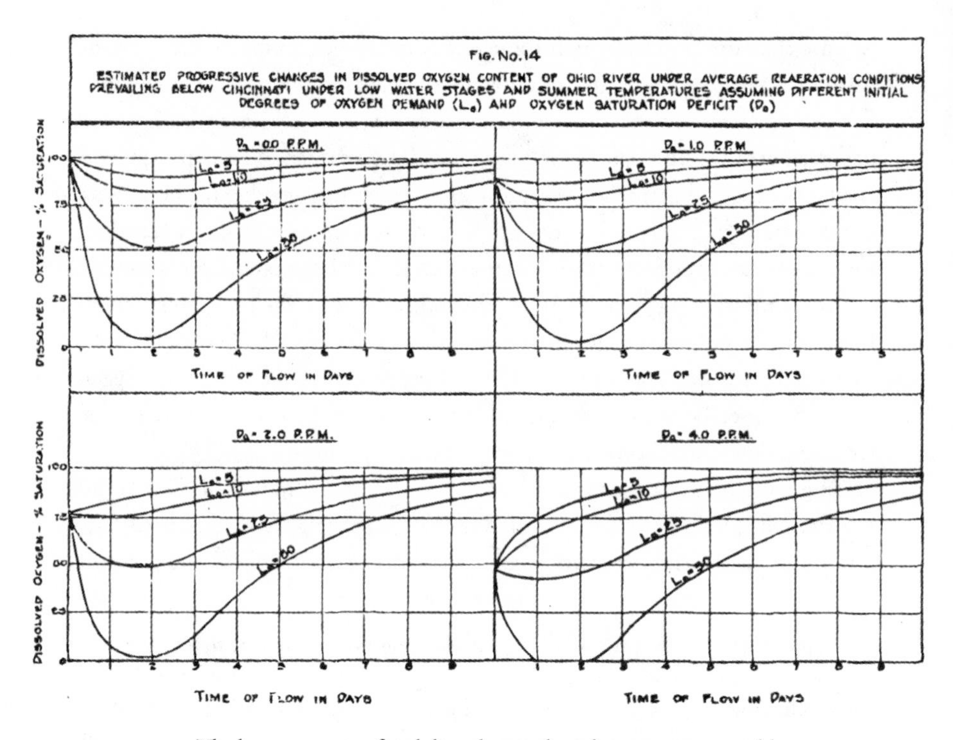

FIGURE 1.1. The bottom curve of each line shows when the oxygen rate could not replenish under different pollution loads (L_a marks different degrees of oxygen demand in relation to higher quantities of organic contamination), and the climb of the line back to the top axis shows the slow regain of oxygen. This is one of several graphs in the report, all with similar curves showing different variables. Image from Streeter and Phelps, *Study of the Pollution and Natural Purification of the Ohio River*, 81.

which increasing time or water velocity does not contribute to an increase in oxygen rates. It is the moment where imbalance occurs and the river chokes. Today, this curve is called an oxygen sag, and the point at the bottom is called the critical point. It is the line that marks the moment of modern environmental pollution. In different places, with different variables, and different amounts of organic pollutant, the numbers and tilt of the curve might change, but the essential sigmoid shape of the curve is the same each time.

Universalism

Based on the consistency and regularity of these results, Streeter and Phelps declared, "It has been shown in the foregoing text that the oxygen self-purification of the Ohio River is a *measurable* phenomenon, governed by *definite laws* and proceeding according to certain fundamental physical and biochemical reactions. Because of the fundamental character of these reactions and laws, it is fairly evident that the principles underlying the phenomenon as a whole are applicable to virtually *all* polluted streams."[36] All streams? What luck! Streeter and Phelps had answered the regulatory wet dream of governing across different places. Now it did not matter if the pollution was of one kind or another, if the water was in a stream or in a tank, in the slow and cooperative Ohio River or the complex and naughty New York Harbour.[37] Legibility across jurisdictions, scales, materials, and contexts is likely a core reason Streeter and Phelps's equation for assimilative capacity was so immediately successful, taken up by regulatory bodies in both theory and practice and hailed as a classic within two decades.[38] Also crucial to its success was the fact that assimilative capacity allowed some dumping of waste to occur. Instead of changing systems that allowed industrial effluents to begin with, governance could turn to technical efforts to locate and manage allowable limits. This is the foundation of the permission-to-pollute system.

The power of "discovering" (read: labouriously crafting) scientific phenomena that hold across bodies of water lies in the Western concept of universalism. Universalism is the claim that "certain principles, concepts, truths, and values are undeniably valid in all times and places and, by extension, the characteristics of phenomena are invariant. Universal knowledge is therefore the opposite of local, particular, and situated. . . . It is transcendental, placeless, and untouched by context."[39] Universalism requires fungibility or "exchangeability. Fungibility

36 Streeter and Phelps, *Study of the Pollution and Natural Purification of the Ohio River*, 59; emphasis added.

37 Many authors have argued that exact, quantifiable definitions and processes are favoured in a policy context, as they lead to fewer discussions and conflicts become solvable. Maarten Hajer (unmarked) argues that quantification is a form of definition that "institutions can handle and for which solutions can be found." Hajer, *Politics of Environmental Discourse*, 15. Also see Verran, "Numbers Performing Nature"; Igo, *Averaged American*; and Bäckstrand, "What Can Nature Withstand?"

38 E.g., Bloodgood, "Water Dilution Factors for Industrial Wastes." A mark of Phelps's success was his placement on the board for the *Report of Committee on Standard Methods of Water Analysis to the Laboratory Section of the American Public Health Association*, the authoritative text for American municipal water workers, which is still in use today. His tenure introduced the first empirically tested thresholds for pollution into the handbook, and they remain there today nearly one hundred years later, tweaked but essentially unchanged. For the most up-to-date version of the century-old handbook, see American Public Health Association, American Water Works Association, and Water Environment Federation, *Standard Methods for the Examination of Water and Wastewater*.

39 That is, the opposite of Land. Castree, Kitchin, and Rogers, "Universalism."

also means getting anatomized into exchangeable parts to be stored, shipped, sold, combined with other parts,"[40] the kind of work that variables do. Streeter and Phelps were scientific universalists, looking for (read: painstakingly creating) universal, fungible traits of self-purification.

Angela Willey (unmarked) writes that most Western sciences, as well as other academic traditions, offer "ultimately totalizing scientific explanations of the world and our place in it [marked . . .] by an implicitly Judeo-Christian brand of secularism that allows us to imagine nature as law-governed." She continues, describing this approach as characterized "by a Eurocentric protomodernity that separates the rational from the irrational,"[41] separates (some) humans from Nature,[42] and separates variables from the background noise of Nature.

The universal is never universal, but rather an argument to imperialistically expand a particular worldview as *the* worldview. Comparative philosopher Mary Graham (Kombumerri, Gold Coast) summarizes these epistemological and ontological beliefs when she writes:

> For most Westerners reality is what it is irrespective of what humans think or know about it; secondly, that reality is ordered, that it has a structure that is universal and invariant across time and place. They claim that the structure and forces of the natural world remain the same in different times and in different contexts. They also believe that this structure is knowable and that Western science has provided the ability to explain, predict and control many natural phenomena and to invent technologies to solve human problems.[43]

This knowledge system provides a top-down, all-encompassing view that feminist Donna Haraway (unmarked) has called the god trick, "seeing everything

40 paperson, *A Third University Is Possible*, 13.

41 Willey, *Undoing Monogamy*, 993. As you can see from the citation, the politics of universality, particularly their spread and enforcement via colonialism, has wide-reaching effects from pollution to sexuality. Though it is not written from an anticolonial perspective, for a history of universalism and science, including its relationship to civilized civilizations, nationality, Christendom, institutional and competitive internationalism, and modes of governance, see Somsen, "History of Universalism." White supremacy and racism are not covered in the article, though it does cover World War II and German nationalism, including Nazis and genocide. Weird, right? As an antidote, the following text is very good on linking the role of universalism to race, racism, and Nature: Moore, Kosek, and Pandian, *Race, Nature, and the Politics of Difference*.

42 Moore, Kosek, and Pandian, *Race, Nature, and the Politics of Difference*.

43 Graham, "Understanding Human Agency in Terms of Place," 71.

from nowhere."[44] It is unattached, unaccountable. No wonder Nature was born from this worldview.

Since the Enlightenment, a goal of universal science has been what science historian Lorraine Daston (unmarked) calls "European self-portraiture," particularly as the borders of Europe became extended and even ambiguous during (ongoing) imperial and colonial conquests—the goal was to make mini-Europes and Europeans through science.[45] Imperialism and colonialism both involve the scientific appropriation of local and Indigenous knowledges, eaten up and digested to create dominant scientific knowledge.[46] Historically, this included botany and the cultivation of economically valuable plants,[47] an interest in climate[48] and the expansion of agriculture,[49] and the control of diseases such as malaria,[50] all of which enabled successful settlement.

Simultaneously, the dominant scientific knowledge system was and is of-

44 Haraway, "Situated Knowledges," 581.

45 Daston, "History of Science." Though for more on how the making of European subjects via science never quite worked for colonized subjects, see Seth, "Putting Knowledge in Its Place." Thank you, Suman (unmarked), for your work, but also for your candor, generosity (especially knowing and taking the time to introduce me to local water), and collegiality.

46 The legacy of dominant and imperial science eating up and getting fat off of local and Indigenous knowledge is fashionable again today in the grant-supported drive for Traditional Ecological Knowledge (TEK), Traditional Knowledge (TK), and Indigenous Knowledge (IK). While dominant science's aims are often articulated as a drive toward inclusion (into Empire, I assume), its attempts to "incorporate" (use, assimilate, ingest, nom nom nom) Indigenous knowledges are often another form of colonialism that extends the reach of colonial and settler goals by acquiring more types of data. Most scientists and staff at federal granting agencies I work with do not appear to understand that TEK et al. are about ways of knowing, not what is known. Ain't no data in TEK. This is the main reason CLEAR *does not* claim to engage in TEK knowledge collection (ew), integration (ow!), or use (ugh). This doesn't mean that some of our Indigenous members and colleagues don't use their own diverse knowledges to get scientific work done—it means we don't make that legible or available for munchies. More of this in chapter 3. For articles on the perils of inclusion of TEK, etc., see McGregor, "Traditional Ecological Knowledge"; Reo, "Importance of Belief Systems in Traditional Ecological Knowledge Initiatives"; Nadasdy, "Politics of TEK"; Nadasdy, "Anti-Politics of TEK"; and Dene Nation and Assembly of First Nations, "We Have Always Been Here."

47 Schiebinger and Swan, *Colonial Botany*; Brockway, "Science and Colonial Expansion."

48. Zilberstein, *Temperate Empire*.

49 Knobloch, *Culture of Wilderness*.

50 Fanon, "Medicine and Colonialism"; Nash, *Inescapable Ecologies*; Bashford, "'Is White Australia Possible?'"; Lyons, *Colonial Disease*.

ten articulated as a reason for colonialism: "The idea that science and technology were among the gifts that Western imperial powers brought to their colonies was an integral part of the discourse of the 'civilizing mission,' one vaunted by both proponents and critics of the methods of colonialism."[51] Christianity, residential schools, and dominant science were different techniques through which colonizers claimed to bring light to the darkness of primitivism while simultaneously maintaining a difference between the colonized and colonizers.

Feminist scholar Helen Verran (white [oyinbo]) describes the "abhorrent moral economy" of teaching dominant science and mathematics to colonized peoples, or even just well-intentioned comparative analyses of different knowledge systems. She writes how Yoruba counting and knowledge "could only be taken as an echo, a shadowy form of English logic. The schema reenacts the categories of a universal modernity, originating in European traditions, and a Yoruba echo of a necessarily European modernity. Either way, a distinct 'us' and 'them' are locked forever together, and apart, through the spectre of originality/mimicry. . . . The only way to tell such difference is to pull 'their' world into 'ours.'"[52] This kind of access—being able to pull entire worlds into dominant worlds—is a hallmark of colonialism.

In short, dominant science can be used to fuel a militant universalism where a single knowledge becomes the touchstone for all other knowledge systems, which either can dismiss and erase other forms of knowledge or can place those knowledges in the waiting room of modernity[53] as late, quaint, cute, curious, undeveloped, and consumable for settler desires, well intentioned or not. This is not just a historical problem.

51 Seth, "Putting Knowledge in Its Place," 373. This discourse is alive and well today. STEM camps for Indigenous youth are a mainstay of grants and (settler) institutional celebrations at my university, where the assumption is that STEM will lift up, enable, empower, and otherwise better Indigenous youth. It remains a civilizing force, a force to bring "them" in line with "us," a way to diversify empire. It is a constant battle to remind people that Indigenous people already have robust ways of knowing, in place, and using scientific degrees or university enrollments of Indigenous people is only a metric of success within a colonial logic. See Megan Bang's work for a different take; e.g., Bang and Medin, "Cultural Processes in Science Education."

52 Verran, *Science and an African Logic*, 31. This text is excellent in terms of showing a sustained engagement in the tensions between universalism and specificity in counting, which is often assumed to be the most universal act of STEM knowledge.

53 Chakrabarty, *Provincializing Europe*.

There's a story we tell in the lab about the academic peer review of our first cod paper.[54] *Since we collected guts from fishers, the reviewer wants to know: "how did the authors know they were cod?" I remember reading the question out loud during the lab meeting and people laughing. Of course, it was cod—the fishers said so! Cod is a major part of the culture, heritage, diets, livelihoods, songs, and life for the settlers on the island of Newfoundland that we got the cod from. Babies know what cod is! Since we could not write "because Newfoundland" in response to the reviewer and still be published, we assured them that a lab member was present during gutting. More laughing. Sometimes, that person was our newest lab member, Alex, who has the most scientific training of the group but is also from landlocked and cod-less Saskatchewan. Someone in the lab jokes, "He couldn't tell a goldfish from a mackerel." Alex is a good sport—he mimes confusion. The reviewer accepts our explanation: There was a scientist present. The paper is published.*

One Pollution, One Nature

So Many Pollutions

Before he went to the Ohio River, Phelps wrote, "Of immediate and pressing interest is the fixing of standards of permissible pollution, which will comply with the common law conception of reasonable use and develop the maximum advantageous use of the streams."[55] It's not that Phelps was a jerk—on the contrary, he might be considered a proto-conservationist who sought to find the maximum use of a waterway *without* abandoning it entirely to pollution, the common practice at the time.[56] Phelps was working in an environment where the pressing scientific, public health, and governance question was to find the demonstrative difference between polluted and safe drinking water. It was a big problem. In 1868, commissioners in Britain who had been appointed to craft what would become the Rivers Pollution Prevention Act wrote, "There is no such thing as absolutely pure water in nature, and the waters met with in our springs, lakes, rivers, and sewers, form a series gradually increasing in dirtiness;

54 The published paper in question is M. Liboiron et al., "Low Plastic Ingestion Rate in Atlantic Cod (*Gadus morhua*)." This is also the paper that used the guts that Charlie tells the story about earlier in this chapter.

55 Phelps, "Stream Pollution," 928.

56 Tarr, "Industrial Wastes and Public Health." Also see Phelps, "Discussion."

there is actually no definite line of demarcation separating the purest spring water from the filthiest sewage. . . . It is, therefore, obvious that, for the purposes of efficient legislation, an arbitrary line must be drawn between waters which are to be deemed polluting and [those deemed nonpolluting]."[57]

Before Streeter and Phelps made that line less arbitrary, definitions of environmental pollution proliferated. Some were chemical, focusing on traceable aspects of sewage: "[Polluted water is] any liquid containing, in solution, more than two parts by weight of organic carbon, or 3 parts by weight of organic nitrogen in 100,000 parts by weight."[58] Even with the advent of the germ theory, there was no bright white line to demark exactly how many germs would constitute unsafe water: "[*B. coli*'s] presence in water is to some extent indicative of pollution, although its abundance rather than its mere presence must be considered as the criterion."[59] But what abundance? Other definitions of pollution were aesthetic or available to lay evaluation: "A 'relatively purified' river: Should no longer make the stream slimy or muddy, should contain neither the remains of beetroot or cosettes"[60] and should be "tasteless and inodorous . . . and is incapable of putrefaction, even when kept for some time in closed vessels at a summer temperature."[61] The range of analysis was dizzying—not what a centralized government wanted.[62] The adoption of the threshold theory of pollution that the Streeter-Phelps equation promised allowed regulatory bodies to replace these varied definitions with a single one based on assimilative capacity.

57 Glen, "Appendix B," 75.

58 Bailey-Denton, "Sewage Disposal," 9.

59 American Public Health Association et al., *Standard Methods for the Examination of Water and Sewage*, 84. While a 1914 water-quality standard for interstate trains—the first in the United States—provides a maximum number of bacteria in a sample (100 per cubic centimetre), it also states that in "the attempt to establish limits of this kind it [is] inevitable that manifold difficulties should have been encountered[, including] the difficulty inherent in any attempt to establish an exact line of demarcation between two such extremes as undoubtedly safe water supplies and those which should assuredly be condemned." Monfort, "Special Water Standard," 66, 69. This sentence disappeared from the text after Phelps joined the committee that wrote the *Standard Methods.*

60 Naylor, *Trades Waste*, 5–6. I have been unable to identify what a cosette is, though I look in my water regularly, ever hopeful.

61 Massachusetts State Board of Health, *Seventh Annual Report*, 26.

62 In addition to writing about the history of water chemistry and analysis, Christopher Hamlin (unmarked) has also written about another area of universalization—that of water. Water also used to be a lot of different things. Hamlin, "'Waters' or 'Water'?"

Managerial Ontologies

The threshold—that arbitrary line between pollution and nonpollution, made less arbitrary through oxygen sag—is where policy, accountability, and responsibility come together.[63] When Phelps laid out his theory of assimilative capacity in 1912, he argued that, once the self-purification rate of a river was determined,

> the ultimate oxygen requirement of the sewage of a community expressed in pounds per day may then be balanced against the available oxygen resources of the stream into which this sewage is to be discharged, and the resulting figure will show the effect of such discharge upon the dissolved oxygen of the stream. If the effect is to reduce the dissolved oxygen of the stream below the permissible point, further purification [by municipal systems] is indicated. If two or more communities are contributing to the same body of water the total oxygen of the water may be apportioned in an equitable manner between them.[64]

The logical extension of quantifying the threshold of pollution was to parcel out assimilative capacity—essentially, the ability to waste, even the right to waste—to polluters. The area under the curve became a sacrifice zone,[65] designed for pollution.

Managing the line between contamination and pollution, now differentiated as fundamentally different states, became a defining feature of environmental governance, and remains so today.[66] The first public standards for drinking water in the United States included threshold "concentration limits for lead, fluoride, arsenic, and selenium."[67] Writing about the history of toxics regulation,

63 For brilliant work on this concept from a present-day, European pollution governance perspective, see Olsson, "Setting Limits in Nature and the Metabolism of Knowledge."

64 Phelps, "Chemical Measure of Stream Pollution," 534.

65 Lerner, *Sacrifice Zones*.

66 The way that assimilative capacity, natural thresholds, and the ability to pollute to a specific level have combined to become a defining feature of current-day environmental regulation is covered by excellent research in the natural sciences, environmental management, social sciences, and history, including Busch, "Use and Abuse of Natural Water Systems"; Walker, *Permissible Dose*; Olsson, "Setting Limits in Nature and the Metabolism of Knowledge"; Lueck et al., "Determination of Stream Purification Capacity"; Sayre, "Genesis, History, and Limits of Carrying Capacity"; Firth, "Status of Water Quality Modeling in the Pulp and Paper Industry"; Schneider, *Hybrid Nature*; and Hajer, *Politics of Environmental Discourse*.

67 United States Public Health Service, "Public Health Service Drinking Water Standards," 373.

environmental historian Frederick Davis (unmarked) writes that the role of the United States' federal Environmental Protection Agency "is to implement the pollution control laws enacted by Congress. Its most important function in this respect is to establish National standards that govern *how much pollution is allowable*."[68] Earlier periods' flexibility and diversity of what counted as pollution diminished to the point of irrelevance[69] from a policy and scientific standpoint. Pollution became—is—assimilative capacity.

Thresholds Abound

Streeter and Phelps did not develop their ideas in a vacuum. Other sigmoid (*S*-shaped) curves that showed threshold moments were being produced in the early decades of the twentieth century across scales, nations, and disciplines. For example, Aldo Leopold (unmarked), the famed American environmentalist, used the concept of carrying capacity to manage wildlife areas in the early 1930s, using a sigmoid curve to show how populations would level off.[70] He then designed landscapes to carry a maximum load of animals.

Leopold's large-scale field experiments had already been anticipated by "laboratory studies of fruit flies, flour beetles, or other convenient organisms"[71] raised, studied, and extinguished in glass vials. Their lives and deaths produced *S*-shaped population curves, as did the population data of colonized Algeria, a

68 F. Davis, *Banned*, 2; emphasis added.

69 Since Streeter and Phelps, there have been ongoing critiques and pushback on the concept of assimilative capacity from *within* the sciences. The list of these critiques (below) used to contain two pieces that have been removed, which were two of the works most aligned with my own historicization of how certain toxicological truisms were built in specific ways. But it turns out the author is a litigious thinker dedicated to undoing all forms of environmental monitoring, threshold and otherwise, in the service of polluting industries. I point this out for two reasons. First, I am making absent citations conspicuous as part of a politics of citation. Since citation is a reproductive technology, a way to build and rebuild fields of knowledge, and a form of action, I want to both omit and mark that omission. Second, I want to highlight how data and analysis never speak for themselves. The same data or critique can be used to further opposing concepts of what is good, right, and true. Industry has activists, too. Susan Sontag (unmarked) makes this point for evidentiary photography in *Against Interpretation*. For citable work that critiques assimilative capacity from a scientific perspective, see Busch, "Use and Abuse of Natural Water Systems"; Campbell, "Critique of Assimilative Capacity"; Cairns, "Assimilative Capacity Revisited"; Cairns, "Threshold Problem in Ecotoxicology"; Westman, "Some Basic Issues in Water Pollution Control Legislation"; and O'Brien, "Being a Scientist Means Taking Sides."

70 Leopold, *Game Management*.

71 Odum, *Fundamentals of Ecology*, 123.

"natural experiment" on colonized not-quite-humans.[72] Before the term *assimilative capacity* was used to describe organic waste in water, a version of the term was used in nutrition science to describe thresholds in the absorptive powers of the body, "used to designate broadly the ability of the organism to convert the digested nutrients of the feed into body tissue."[73] The scale of the body had long been associated with thresholds for harm—this is the very premise of toxicology since Paracelsus (unmarked) famously stated, "What is there that is not poison? All things are poison and nothing is without poison. Solely the dose determines that a thing is not a poison."[74] By the 1920s and 1930s, when Streeter and Phelps were conducting their work, pharmacology treated sigmoid curves as normal and natural, and toxicological debates pivoted on how those curves should be interpreted and how drugs should be managed, not on the theory of the threshold itself.[75]

By the early twentieth century, the terms that described thresholds, including *assimilative capacity*, *critical load*, *tolerance dose*, *permissible dose*, and *carrying capacity*,[76] were appearing with increased frequency across new and established sciences, applied to scales from cells to landscapes. These *S*-shaped curves offered a density of evidence that harm could be defined as strain that surpassed tolerable limits, rather than a symptom, a discontinuity in aesthetics, a moral dilemma, or another less quantifiable phenomenon. The ubiquity of *S*-shaped curves across different geographies, different bodies, different toxicants, and different scales was interpreted to mean that threshold relations were characteristic of Nature itself, that is, universal.

72 Pearl, *Biology of Population Growth*. One of the most famous of these human population curves is Meadows, Randers, and Behrens, "Limits to Growth." For more on how these curves shaped the discipline of ecology, see Kingsland, *Evolution of American Ecology, 1890–2000*. For more on the politics of population statistics, see Murphy, *Economization of Life*.

73 Armsby, *Nutrition of Farm Animals*, 441.

74 Quoted in Grandjean, "Paracelsus Revisited," 126. For more on the body as a sink for pollution, see Agard-Jones, "Bodies in the System"; Brown, "Last Sink"; and Cram, "Becoming Jane."

75 E.g., Clark, *Mode of Action of Drugs on Cells*.

76 The term *carrying capacity* has a history starting in the 1840s in reference to calculating loads that international ships could carry, before it branched into biology, ecology, and statistics. An excellent history of the term is Sayre, "Genesis, History, and Limits of Carrying Capacity."

Making a Nature Robust within Limits

Just as there were competing definitions of pollution before assimilative capacity, there were also competing definitions of Nature. In *Risk and Blame*, anthropologist Mary Douglas (unmarked) recounts four different cultural views of nature: nature as robust (nature can handle any human intervention), nature as robust within limits (nature can handle human intervention up to a certain point), nature as fragile (nature cannot handle human intervention), and nature as capricious (nature will be unpredictable in the face of human intervention).[77] Streeter and Phelps, with many others, appeared to have provided empirical evidence that nature is indeed robust within limits.[78] They argued that these limits could be precisely located and managed. Pollution became a threshold at the same time that Nature became robust within limits.

Historian Wiebe Bijker (unmarked) might have been writing about pollution and Nature when he wrote that scientific "consensus means that the interpretative flexibility of, for example, an observation statement disappears, and from then on only one interpretation is accepted by all. Such a closure is not gratuitous but has far-reaching consequences: it restructures the participants' world."[79] In this case, one out of many versions of Nature and pollution was naturalized. On the bright side, Bijker writes, "It is in principle always possible—although in practice very difficult—to reopen up a controversy once closure is reached."[80] That's what CLEAR is here for.

Canada is gearing up to make Big Plastic Plans at the federal level. I'm on one of many calls organized by the federal government with plastic pollution

77 Douglas, *Risk and Blame*, 263–64.

78 I cannot overstate the success of this version of Nature. An extreme but quintessential manifestation of the idea that Nature can handle a certain amount of pollution and that this is the proper role for Nature is the story of James Lovelock (yes, the Gaia hypothesis guy), who argued that Gaia (a.k.a. the planet) wanted or at least welcomed a certain amount of pollution to help with maintaining planetary bias. Lovelock published these ideas while under the employ of ExxonMobil (yes, the oil guys). In her exploration of this fascinating story, historian Leah Aronowsky (unmarked) argues that this concept of natural assimilation is the conceptual premise of climate change denial. Aronowsky, "Gas Guzzling Gaia." See also an amazing ad that Aronowsky uses to frame the story, where Mobil argues, "Our point is that nature, over the millennia, has learned to cope. Mother Nature is pretty successful in taking on human nature": ExxonMobil, "The Sky Is Not Falling."

79 Bijker, *Of Bicycles, Bakelites, and Bulbs*, 85.

80 Bijker, *Of Bicycles, Bakelites, and Bulbs*, 85.

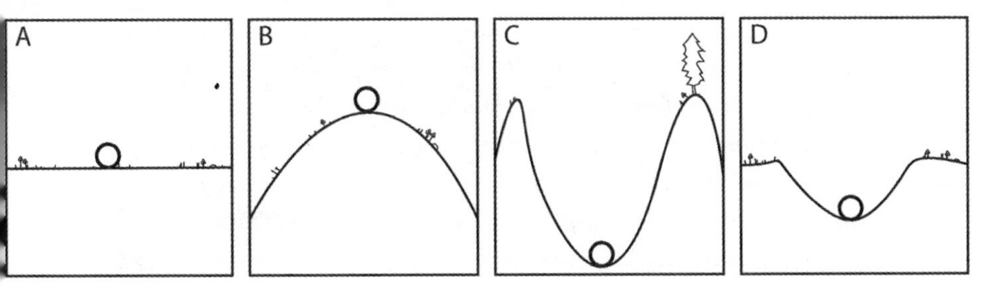

FIGURE 1.2. The four myths of Nature. (A) Nature is capricious: the ball can roll anywhere, anytime. (B) Nature is fragile: the ball can roll off at any moment! (C) Nature is robust: that ball is not going anywhere. (D) Nature is robust within limits: the ball will roll out if we push too hard. Illustration by Max Liboiron. CC-BY 3.0.

scientists from across the country. We're talking about how there are no established standards for analysing and reporting marine plastic research and how this makes studies difficult to compare and validate, a reoccurring discussion in the field. Someone brings up how we should establish a threshold for plastics. I startle and mash on the buttons to unmute my mic, accidentally hanging up on the call. I call back faster than I've ever called into a conference call before and exclaim, "Plastics are not a threshold pollutant!"[81] *The other scientists and feds hear me out, and in the end agree that maybe we don't want to articulate a threshold,*[82] *since "industry" (which includes the*

81 This is not the same as exclaiming, "Fuck thresholds and the horse they rode in on!" Saying that plastics are not a threshold pollutant is a scientific statement that aligns plastics with other pollutants like radiation or carcinogens where there is no "safe dose." I do not always say what I mean when doing political work as a scientist—I often say things adjacent to what I mean so that I can remain legible to my audience. The threshold theory of pollution is not really up for debate, and discussions of universalism, thresholds, and colonialism simply won't make sense on a federal conference call. This is one way to talk about compromise, which will be covered more in chapter 3: activism happens on a terrain that is already laid out, has already been identified as unjust and in need of change, and you have to make some sort of sense in that terrain to change it.

As a side note, even a statement about plastics not having thresholds is not guaranteed to circumvent a threshold theory of harm. Nonthreshold pollutants have been given thresholds via risk theory, where a specific amount of death or morbidity is an acceptable loss. The threshold theory of harm is strong and beloved. See the excellent work of Vogel, *Is It Safe?*; Cram, "Becoming Jane"; Walker, *Permissible Dose*; and Langston, *Toxic Bodies*.

82 Actually, there is an articulated threshold for plastics, of sorts. It's for Northern Fulmar, but it occasionally leaks into other contexts. It's called the Ecological Quality Objective (EcoQO) measure, and its threshold is 0.1 g / 10 percent of individual birds within a

feds on the call, in my view) can use it as a permissible level to pollute up to. Other callers agree or stay silent. We move on to a different universalizing topic about the one best way to report microplastic findings. Small victories.

Update one year later: a Canadian federal funding call for research on plastics prioritizes research on "the biological, chemical or physical stressors introduced to ecosystem components by these contaminants to identify thresholds needed to disrupt biochemical, physiological or behavioral interactions between plastic, the environment, and organisms."[83] *The threshold theory of harm is formidable in its resilience.*

Resource

When Nature becomes robust within limits and threshold theories of harm are dominant, land relations become managerial rather than reciprocal. In colonial understandings of Nature, (certain) humans can protect, extend, augment, better, use, preserve, destroy, interrupt, and/or capitalize on robust-within-limits Nature. That is, Land becomes a Resource. Resources refer to unidirectional relations where aspects of land are useful to particular (here, settler and colonial) ends. In this unidirectional relation, value flows in one direction, from the Resource to the user, rather than being reciprocal as legal scholar Andrew Brighten (unmarked) notes in his interpretation of court proceedings on xʷməθkʷəy̓əm relations to salmon:

> The court quickly moves past a scant two sentences referencing the Musqueam social ontology of salmon and humans bonded in a reciprocal relationship, then distills this relationship to the activity of "taking"—

population. Basically, 0.1 grams of plastic ingested by more than 10 percent of birds studied is too much. Scientists note, "This [EcoQO] target had no substantiated background of ecological or individual or population health. It represent[s] an arbitrary target considered to reflect 'acceptable ecological quality' as used in policy documents." This threshold does not come from Nature, or even from science, but from policy documents for the purpose of policy. Regardless of whether we want to argue that this is a good or bad measure, it shows just how strong the threshold logic is within dominant science and how it is *the* guiding logic in environmental governance within the colonial state. Provencher et al., "Quantifying Ingested Debris in Marine Megafauna," 1467. For more on the articulation of EcoQO as a threshold for policy rather than for harm, see the measure in action in Van Franeker et al., "'Save the North Sea.'"

83 Government of Canada, "Increasing Knowledge on Plastic Pollution Initiative."

directly opposed to the Musqueam understanding of being given—salmon for food, social and unspecified "ceremonial purposes." The court then repeatedly characterizes Musqueam interaction with salmon as participation in an "economically valuable" "natural resource," [and] "recognizes" the desires of "numerous interveners representing commercial fishing interests." . . . This "resource management" mindset is not unique [to this court case].[84]

This passage documents the flattening of Land relations into Resource relations. In a colonial worldview, a Resource relation is good and right. A body of excellent scholarship critiques this notion of Resource as a colonial, settler, and imperial concept.[85] Building from this work, I will specifically focus on how the colonial logics of Resource are reproduced in practices and concepts of modern environmental pollution.

Standing Reserve

In "The Question concerning Technology," philosopher Martin Heidegger (unmarked)[86] describes a Resource-based arrangement of relations as a "standing re-

84 Brighten, "Aboriginal Peoples and the Welfare of Animal Persons," 39. Anglicisation of xʷməθkʷəy̓əm in the original.

85 For more on conflicting settler-state and Indigenous worldviews over the concept of resource, see Carroll, *Roots of Our Renewal*; Nadasdy, *Hunters and Bureaucrats*; and Reo and Whyte, "Hunting and Morality as Elements of Traditional Ecological Knowledge."

86 As you read what follows, you may notice that Heidegger is the first white man whose theories I've engaged with at length. He was also a Nazi. Why cite him and pick up his theory if my citational politics are about reproducing good relations in knowledge production? Good question. It's not because "who else could I possibly cite?" I cite Heidegger to demonstrate that even anti-Semitic, white supremacist, Nazi, canonized European thinkers not only are well aware of colonial land relations but also can see them with great clarity and nuance. I do this to argue against theories of change that rely on awareness as the crux of motivation for change. "If only settlers and colonialists *knew* how their worldview centers themselves and their needs against others'! If only they were *aware* of their privilege. If only they *understood* their own colonial relations with the world." Heidegger shows: they understand. Paraphrasing Eve Tuck: "Ask ourselves: what if settlers knowing didn't change anything? What if we didn't wait for others to know and were inspired by our own knowing?" This is why I quote Heidegger. And now we never have to deal with him again.

For a prolonged analysis that reads Heidegger and Vine Deloria Jr. (Standing Rock Sioux) together and in the end finds that Deloria does everything Heidegger does but better, with more flair, and without being tied to Nature (or Nazis), see Duarte, *Network*

serve."[87] He writes that creating a resource begins with "enframing," an act where "the energy concealed in nature is unlocked, what is unlocked is transformed, what is transformed is stored up, what is stored up is, in turn, distributed, and what is distributed is switched about ever anew. Unlocking, transforming, storing, distributing, and switching about are ways of revealing. But the revealing never simply comes to an end. . . . Everywhere, everything is ordered to stand by, to be immediately at hand, indeed to stand there just so that it may be on call for further ordering."[88] Heidegger calls the constantly deferred result of enframing the creation of a "standing reserve." He argues that modern technology's main task is to transform and stockpile Nature as a standing reserve via enframing. To paraphrase, this process makes the various relations of Land into a unidirectional relation called Resource for anticipated settler use. He illustrates his theory with the example of coal mining: coal is dug out of the ground not to be present as coal, but to be stockpiled and used at an unforeseen but anticipated occasion.[89]

The transformation of Land into Resource is achieved not only through the

---

Sovereignty; and Tuck, "Research on Our Own Terms." For notes on Tuck's lecture, which was not recorded, see M. Liboiron, "'Research on Our Own Terms' by Eve Tuck." For more on some problems with theories of change based on awareness, see M. Liboiron, "Against Awareness, for Scale."

87 Heidegger, *Question concerning Technology*.

88 Heidegger, *Question concerning Technology*, 16–17.

89 Heidegger, *Question concerning Technology*, 15. In case you are a Marxist geek (hello!), *standing reserve* is a model of constant circulation and deferment, which is also the role of capital in capitalism. In *Capital*, Karl Marx (unmarked) expresses the basic activity of capitalism in the formula M-C-M′ (money [M] buys materials to make a commodity [C] to be sold to make more money [M′]). It is a form whose end is to procure surplus and nothing but surplus. The commodity, whether a manufactured item or a sink for pollution (which we will discuss in a moment), is merely a placeholder and transitory state for increased value for unspecified but eagerly anticipated ends (M-C-M′-C-M″-C-M‴). Marx, like Heidegger, finds the material perfection of this activity in modern technology. He writes, "In the large-scale production created by machines, any relationship of the product to the direct requirements of the producer disappears, as does any immediate use value. The form of production and the circumstances in which production takes place are so arranged that it is only produced as a vehicle for value, its use value being only a condition for this." Marx, *Capital*, 374. Capitalism and colonialism make such friendly bedfellows in part because of these analogous relational logics. Whether it is coal in the form of a standing reserve, a commodity as a placeholder for profit, or assimilative capacity as a sink for future pollution, the model of capitalism and imperialism is one of incessant increase and expansion for settler futures, both of which require access to Land.

arrangement of space but also through the arrangement of time. The temporality of Resource is anticipatory—it makes and even aims to guarantee colonial futures.[90] Crucial to this temporality is the belief that this future can be chosen and that the present can be directed toward it via management practices.[91] As such, Resources eclipse other possible relations with Land both now and in the future. The future is reserved for settler goals, colonized in advance. The landscape cannot support other relations, activities, or futures that might interfere with future use. If a river is for waste assimilation, it can only be a fishing spot if those activities do not preclude its role as a sink.[92] Fostering some futures and eclipsing others is a key technique of the managerial ontologies that characterize Resources and pollution. Risk management, disaster plans, homeland security (and other securities)[93] all share managerial ontologies dedicated to containing time for chosen futures.

One method for detecting plastics in biota such as mussels is to dissolve them in KOH (potassium hydroxide). One of my students wanted to use it for her thesis—it speeds up the process considerably, since instead of dissecting fish guts and spending hours staring over a sieve, you put the guts in a jar in an incubator and come back a week later to nothing but clear liquid, a bit of fatty residue, and some nice floating plastics. Sure, I said. As I ordered her supplies and went through university protocols for hazardous materials, I began to realize how toxic KOH was. For the first time in CLEAR's history, we

90 For more on this concept of a guaranteed future (a.k.a. security) and how it's a primary framework for state exceptionalism and violence, see Masco, "Interrogating the Threat"; and Masco, *Theater of Operations*. For more on white, settler colonial temporality as a concept within this framework, see Mawani, "Law as Temporality"; Mitchell and Chaudhury, "Worlding beyond 'the' 'End' of 'the World'"; Rifkin, *Beyond Settler Time*; Veracini, *Settler Colonial Present*; and Strakosch and Macoun, "Vanishing Endpoint of Settler Colonialism."

91 The failures of these anticipatory tactics are discussed in such books as Bavington, *Managed Annihilation*; and Scott, *Seeing like a State*.

92 Phelps participated in debates about defining pollution according to the use of the water: "Pure water has been defined as a water that contains no harmful or deleterious substances with respect to the purpose for which it is to be used. In accordance with this very practical definition typhoid germs constitute an impurity and calcium salts do not in the case of a drinking water supply, while the reverse is true in the case of a boiler water." Phelps, "Stream Pollution," 928. What is interesting about these debates is how they deal (and specifically fail to deal) with a hierarchy of uses that implicitly ensures that industry uses are always possible.

93 Masco, *Theater of Operations*.

had to order hazardous waste containers and, when they were full, pay for them to go . . . somewhere. I was surprised by how easy it was to just get rid of the waste—just fill out a form and call a guy. Bam! Gone! (Somewhere!) It seemed antithetical to create hazardous waste as a lab dedicated to mitigating pollution. Worse, it was rude to dissolve our relatives and have them leave the lab as hazardous froth.

When I say that colonialism means ongoing settler access to Land for settler goals, this includes access to futures. Settlers do not have to set foot on the Land, own the Land, or even use the Land as a Resource so long as the Land is available for settler futures. You can just order KOH whenever you want, because the infrastructure anticipates its use and disposal as hazardous waste.[94] You can also choose not to, but Land is still arranged as a standing reserve, just in case you change your mind. In this way, a seemingly simple and certainly common scientific research method that produces hazardous waste is involved in colonial Land relations, even though its users are also likely invested in environmental goods and perhaps see themselves as Indigenous allies—or are Indigenous scientists themselves. Colonialism is not an event, not an intent. It is "not even a structure, but a milieu or active set of relations that we can push on, move around in, and redo from moment to moment."[95]

#env law

Property

Pollution is a property right. In *The Colonial Lives of Property*, Brenna Bhandar (unmarked) describes property ownership as "a bundle of rights that can be rearranged and redistributed depending on the social and political norms that legislators aim to promote. . . . The degree to which each of these rights is protected varies; the 'stringency' with which each of these rights in the bundle, such as the right to use, possess, exclude, devise, alienate, etc., can be understood

94 Based on the KOH issue, I made a new guideline for the lab: no processes that necessitate hazardous waste. It means we cannot study bivalves, crustaceans, and other invertebrates for plastic ingestion since KOH is the only way to "dissect" them. I gladly take up the restriction. Though now that's getting complicated as my Inuit colleagues want to study plastics in bivalves in their traditional food webs. Colonial technologies used for Inuit goals are . . . ? What? Not colonialism, but we still have a problem. To be continued . . .

95 King, *Black Shoals*, 40. For an excellent summary of the problems with White settler studies of colonialism, including Wolfe's perpetually quoted bit that "colonialism is a structure," see the second part of chapter 1 in King, *Black Shoals*.

as existing in a hierarchy whereby some rights . . . are more powerful than others."[96] This hierarchy is codified in most environmental regulations: "It is legal for some pollution to occur under Canadian and U.S. environmental law. Under the permission-to-pollute system in Canada, some effluents can be released to a certain amount, and spills and leaks are considered acceptable risks even though they happen regularly. . . . [This right supersedes] the United Nations Declaration of the Rights of Indigenous People and the right to free, prior and informed consent. This includes consent to be polluted or not."[97]

Under current settler state laws in Canada and the United States, the twinned values of appropriation and possessiveness allow different acts of pollution to make logical and even moral sense. The Foundation for Economic Education, a libertarian think tank, teaches, "If I deliberately pipe sewage into a pond on my own land, presumably I consider using the pond as a cesspool to be its optimum use. Hence, there is no abuse, no pollution. If however, either purposely or inadvertently I allow my sewage to flow into a neighbor's pond, against his [*sic*] will, I am without question polluting. I am lowering the value of his [*sic*] property."[98] In short, land is pollutable (so pollutable it can barely be considered pollution!) because a property owner has designated its pollution a best use, but only to the point that it would not infringe on another landowner's right to appropriation and possession.

In 1919, before he had provided empirical evidence for the threshold theory of pollution, Phelps wrote,

> It is good law as well as good economics that a riparian owner is entitled by right to any proper use of the stream that flows by his land, with due regard to the exercise of a similar right on the part of lower riparian users. This dictum of the common law has been interpreted by the courts in some extreme instances to mean that there shall be no appreciable reduction in the flow or alteration in the quality of the water by any user. Such extreme interpretation, however, would of itself defeat the very purpose of the law by prohibiting almost every valuable use of the water.[99]

For Phelps, use was paramount to defining pollution. In his mind, water was a Natural sink.

Yet specific uses did not matter as much as *access* to water for those uses. That

96 Bhandar, *Colonial Lives of Property*, 19–20. Also see Underkuffler, *Idea of Property*.
97 CLEAR and EDAction, "Pollution Is Colonialism."
98 Cooley, "Pollution and Property."
99 Phelps, "Stream Pollution," 928.

is, access to land for settler goals guided his convictions about pollution and resource. Tuck and McKenzie write, "The most important aim of recasting land as property is to make it ahistorical in order to hack away the narratives that invoke prior claims and thus reaffirm the myth of *terra nullius*,"[100] allowing settlers to think of their uses of land, for pollution or otherwise, as proper and right because it belongs to them and their goals and futures, whether individual or collective.

This is not an abstract claim. I cannot overemphasize how assumed access to land is foundational to so many settler relations. Land relations are central not only to Indigenous worlds, but also to settler worlds. To illustrate this with an example about research itself, the following story is by Lauren Watwood (settler), an anthropology master's student who is using CLEAR as a "field site"[101] for her ethnographic research. The story is from her first full day in the lab:

> *Tuesday morning. I'm sitting in the lab. My mind voraciously cataloguing every interaction, every gesture, every idea discussed as I listen to Max efficiently plow through Natasha, Kaitlyn, and Charlotte's list of items to be discussed. "My, my." I think to myself. "I am doing quite a good job being an anthropologist! Mhhmm. Look at all this gold I've already collected." After logistics are attended to, Max settles down in a chair to my left.*
>
> *"OK, what do we need to talk about?" she asks. I say, "Let's discuss our expectations for what I am doing here at the lab." We chat for a few moments and she innocuously asks, "Have you already been collecting data?" Proud of myself and my anthropological ways, I reply, "Yes, I have!"*
>
> *She replies calmly, deadpan: "That's stealing."*
>
> *My brain goes blank. I can't comprehend what she said. I recognize the words to be English, my mother tongue. . . . Yet, I don't understand what they mean in this context.*
>
> *"That's stealing," Max reiterates, likely repeating her words in response to my utterly vacant face. "You came in assuming entitlement to extract data and acted in a deeply colonial, imperialist manner. You thought you could come in here and take information from us without our consent, even after we've talked about needing to have a consent process in place. That's harm-*

100 Tuck and McKenzie, *Place in Research*, 64.

101 Where CLEAR does most of our research, in Labrador, *fieldwork* or *field site* are dirty words. They imply an outside, a Natural wilderness, a terra nullius ready for scientific discovery by settler academics, when in fact these places are homelands, homes, and houses. CLEAR is also a house and a home. While I have heard many people say this, I have never found someone write about it in a document. If you find something, please send it along. Maarsi.

ful." Her delivery of this news was not overtly aggressive, nor accusatory. She was simply explaining the fact of the matter.

"No!" I think, grasping for words that would make her understand. "No! No, not at all! I'm not stealing! I'm doing research!" I screech in my mind, the words clawing to escape my throat, so Max will understand. Please, understand. I wasn't ready to concede that what I had said and done was wrong or was out of alignment or was unethical. Because I couldn't think straight. I felt like I was being attacked and was terrified and pissed and defensive and upset.

Later, I realize: I claimed what wasn't mine to claim and never once questioned my methods. That is colonialism.[102]

To her credit, Lauren looked composed during this exchange, even if her eyes were a little big. When I told her she had to apologize to the lab and see what the lab collective wanted to do with the data she had already collected, she went off and crafted an apology,[103] presented it at a lab meeting a couple of hours later, and was welcomed into the lab. Dominant science, and research in general, plays multiple roles in colonial practices of settler access to Land under the logics of property and Resource.

Streeter and Phelps's work was paramount in abstracting Land into quantified and codified entities like assimilative capacity that could then be used to regulate industry's access to Land for effluents so rivers could be used—but not overused—to their fullest extent. Their scientific contribution was to coordinate access, not question it. This propertied, colonial orientation to research continues today, as Lauren discovered.[104]

102 Thank you, Lauren, for allowing me to share this story. As we've discussed in the lab, great, big, charismatic, obvious, event-based, well-witnessed mistakes are gifts to the lab if we work through them. Others in the lab have made similar choices, but they might have been harder to recognize. Your story gives us collective insights into the ways colonialism orients us to relations, regardless of intent.

103 Apology is a key relational method in feminist and anticolonial science IMHO. CLEAR's lab book has a section on apologies, since they're needed in any space where we're trying to unlearn colonial ways of knowing and after a few events that required apology, it turns out no one knew how to do a good job of it. The most up-to-date CLEAR lab book is on our website: https://civiclaboratory.nl.

104 Today is March 23, 2020: I am working from home in self-isolation during a COVID-19 state of emergency in my province (and the world). Researchers continue to go to campus to do research, even though we have shut down labs, asked all personnel that is not critical to go home, and cancelled all "fieldwork" and travel. Still, the entitlement to access space for research and settler goals is strong, even in the face of a collective pandemic. It's amazing/what-the-actual-fuck.

Maximum Use

Streeter didn't advocate for just any kind of Resource relation—he advocated for "*maximum* advantageous use of the streams."[105] Building on the work of Streeter and Phelps, in 1950 the chairman of the Department of Public Health Statistics at the University of Michigan, C. J. Velz, wrote a treatise on the importance of precisely calculating assimilative capacity so as "to take *fullest* advantage of the inherent resources available" in rivers as sinks. His concern was that "natural purification capacity is not being *fully* utilized,"[106] particularly during seasonal events that swelled rivers with extra water. In response, he perfected the sigmoid curve introduced by Streeter and Phelps. No water was to be wasted that was not adding to assimilative capacity!

Economic geographer Morgan Robertson (unmarked) writes that such mathematics were part of "creating a world in which we [settlers] see ourselves as utility-maximising and self-interested, or of rendering the entirety of the biophysical world as classifications and functions, [which occurs] through rather mundane and incomplete acts of reduction and simplification."[107] Extending this, anthropologist Elizabeth Povinelli (unmarked) writes about how colonists encounter forms of life (including L/land) that are not organized on the basis of market values as irrelevant, as irrational, and even as security risks.[108] Today, the logics, techniques, and infrastructures (in forms from pipelines to policy) of maximum use of sinks uphold land as something that is not only pollutable, but properly so.

The Morality of Property

Maximum use has a morality. Using a Resource to its maximum potential is good; to squander it is bad. Philosopher John Locke (unmarked) says so: "Land that is left wholly to Nature, that hath no improvement of Pasturage, Tillage, or Planting, is called, as indeed it is, wast [*sic*]; and we shall find the benefit of it amounts to little more than nothing."[109] A lot of Europeans were into Locke, and the legacy of those ideas is strong today.

In his work on the British privatization of common land and its relationships

105 Phelps, "Stream Pollution," 928; emphasis added.

106 Velz, "Utilization of Natural Purification Capacity in Sewage and Industrial Waste Disposal," 1608; emphasis added.

107 Robertson, "Measurement and Alienation," 397.

108 Povinelli, *Economies of Abandonment*.

109 Locke, *Two Treatises of Government*, 40.

to concepts of waste and wasting, geographer Jesse Goldstein (unmarked) outlines how land that was being used as a commons but not as a Resource became "a landscape of wasted potential" that specifically wasted "the improvers' economic right—presented as a natural right—to realize the maximum productive potential of all things, at all times, and in all ways."[110] When British peasants "failed" to extract maximum value from a shared landscape, they were removed, the land was enclosed and privatized, and the peasants were reintroduced to the enclosures as wage labourers on newly Resource-rich land. Goldstein argues, "Enclosure was a transformation from one moral conception of value to another."[111] More "than a particular historic technique of land reform in feudal England, and more than a collection of individual acts of theft or an uneven distribution of land and resources," dispossession and enclosure of land as Resource is "a general way of seeing the world" based on "a particular (and persistent) logic of expropriation, produced in and as part of the land itself."[112]

The logic of maximum extraction of value was at work not just in Britain, but also in its colonies. There, the moral imperative to improve land, to rearrange Land into Nature and Nature into Resource, was a primary (though not the only) refrain for dispossessing Indigenous peoples from their Land. Mishuana Goeman (Tonawanda Band of Seneca) contends, "property, as has been argued by Indigenous scholars and their allies, is distinctly a European notion that locks together (pun intended) labor, land, and conquest. Without labor to tame the land, it is closely assigned the designation 'nature' or 'wilderness.'"[113]

110 Goldstein, "Terra Economica," 360, 369. For more on enclosure, see De Angelis, "Marx and Primitive Accumulation."

For Métis, this story of enclosure probably sounds familiar. It's the mode of dispossession used in the scrip system via the Dominion Lands Act of 1879, the largest Land fraud scam in Canadian history. More than just a colonial way of seeing the world, the scrip system was a deliberate strategy of Land theft and genocide that worked in a myriad of ways, including turning Land into property parcels, addressing Métis as individuals rather than as a collective culture, and creating bureaucratic systems of land tenure that were impossible to maneuver legally and ripe for fraud. To compound this theft, surveyors and others dispossessed Métis of Land by stealing the scrip coupons themselves. If you're unfamiliar with the scrip system, especially if you are in Canada, you should know that this story is part of the truth part of truth and reconciliation. Here's a primer: CBC Radio, "From Scrip to Road Allowances"; Tough and McGregor, "'The Rights to the Land May Be Transferred'"; Adese, "'R' Is for Métis"; and Andersen, *Métis*.

111 Goldstein, "Terra Economica," 372.

112 Goldstein, "Terra Economica," 372.

113 Goeman, "Land as Life," 77.

Tiffany Lethabo King (Black) agrees, writing that "within this Lockean formulation, Indigenous subjects who do not labor across the land fail to turn the land into property and thus fail to turn themselves into proper human subjects."[114] Civilization and its opposite become identified with land use aligned with creating standing reserve to procure maximum economic value.

In 1876, an Indian reserve commissioner on Vancouver Island in the region currently known as Canada addressed members of "a Native audience" (Nation unspecified), who were being moved to reserves that were a fraction of the size of their previous Land bases. He explained, "The Land was of no value to you. The trees were of no value to you. The Coal was of no value to you. The white man came he improved the land you can follow his [*sic*] example."[115] This settler commissioner, along with many of his contemporaries, thought "that until Europeans arrived, most of the land was waste, or, where native people were obviously using it, that their uses were inadequate."[116] The virtue of "good use" "functions as a usable property to dispossess Indigenous peoples from the ground of moral value"[117] through an "ideology of improvement that privileges European forms of cultivation as proof of ownership."[118] This is only possible when there is *one* right land relation, accomplished via universalism.

In *A Third University Is Possible*, la paperson writes, "Property law is a settler colonial technology. The weapons that enforce it, the knowledge institutions that legitimize it, the financial institutions that operationalize it, are also technologies. Like all technologies, they evolve and spread,"[119] in this case to pollution via modern environmental sciences.[120]

Plastic's Moral Economy

One of the most popularized studies on plastic pollution is titled "Plastic Waste Inputs from Land into the Ocean."[121] Perhaps you know it better from its claim

114 King, *Black Shoals*, 23.

115 Cited in Cole Harris, *Making Native Space*, 108; capitalization of *Land* in original.

116 Cole Harris, "How Did Colonialism Dispossess?," 170.

117 Moreton-Robinson, *White Possessive*, 176. Also see Nicoll, "Indigenous Sovereignty and the Violence of Perspective."

118 Bhandar, *Colonial Lives of Property*, 10.

119 paperson, *A Third University Is Possible*, 4.

120 What I do not cover here is how pollution is also used as a technique to dispossess, even when logics of cultivation and sinks are not at play. For work in this area, see Johnson, "Fearful Symmetry of Arctic Climate Change"; Perreault, "Dispossession by Accumulation?"; and Stamatopoulou-Robbins, "Uncertain Climate in Risky Times."

121 Jambeck et al., "Plastic Waste Inputs."

that "the majority of plastic enters the ocean . . . from just five rapidly growing economies—China, Indonesia, the Philippines, Thailand, and Vietnam."[122] The media really loved it, and it circulated extensively. The research aimed to estimate the amount of postconsumer plastics (not industrial plastics)[123] entering oceans through a model based on population density along a country's coastline, the amount of plastic waste generated within that country expressed in per capita figures,[124] and the percentage of "mismanaged" waste in that country. In the paper, *mismanaged waste* is defined as "material that is either littered or inadequately disposed. Inadequately disposed waste is not formally managed and includes disposal in dumps or open, uncontrolled landfills, where it is not fully contained . . . plus 2% littering."[125]

It is not clear how the authors of the study compiled these numbers.[126] The

122 Jambeck et al., "Plastic Waste Inputs," 771.

123 Postconsumer waste (waste flows that start after items are purchased in the consumer market) accounts for only one area of marine plastics. Nurdles, or industrial production pellets, are regularly found in the ocean, and the model does not account for industrial and manufacturing-scale waste from production, transport, and construction. There are two problems with this. First, the only positive, intentional, large-scale change in marine plastic pollution trends is the global reduction of nurdles that occurred between the 1970s and today, likely due to industry response to early plastic reports. This means that one of the only available success stories about marine plastics is rarely told. Second, reports like these reproduce the erroneous truism that plastic pollution is a consumer problem rather than an industrial production problem. For more on how this framing is not unique to plastics but rather common to other environmental framings of waste generally, see Lepawsky, *Reassembling Rubbish*. Thank you, Josh (white settler), for your intellectual comradery and the generosity of your careful thinking and simple yet consistent ethics. One of the intellectual joys of working in Newfoundland has been to think with you about waste, especially on the thorny issues of Mary Douglas, scale, and action. Thank you especially for your humility, commitment, and carefulness in these discussions, and knowing where you ought to pick up lessons on your own. It is a privilege to work alongside you.

124 Per capita waste measurements erase both the role of industry in creating disposables as well as inequities in wealth that impact how waste flows through households and regions. They are made by taking the total amount of municipal waste recorded for a region, then dividing that number by the population of the region. They are an intense flattening of difference. For more on the justice problems of per capita waste measurements, see M. Liboiron, "Politics of Measurement."

125 Jambeck et al., "Plastic Waste Inputs," 756, 769.

126 There is no record of the raw data, the types of data, the categories of data, its sources, or its analysis. The report does note that there are several uncertainties in the data, including "relatively few measurements of waste generation, characterization, collection, and disposal, especially outside of urban centers. Even where data was available, methodologies were not

countries with the highest "mismanaged waste" figures are also countries most likely to have high data uncertainty. North Korea, for example, has a 90 percent "mismanagement" rate, a number that may well reflect a lack of public reporting practices rather than waste practices. Regardless, the category of mismanaged waste includes dumps (rather than landfills), informal recycling (rather than municipal recycling), reuse and repair sectors, and waste that is not managed at all: basically, deviations from American styles of land use and waste management. Another way to interpret the figures is that the countries listed are those that least resemble the United States. Given these assumptions, it is not surprising that the model's projections have not held up. When they were tested in the Pearl River Delta in China, they overestimated plastic waste flowing into the marine environment by an order of magnitude.[127]

---

always consistent, and some activities were not accounted for, such as illegal dumping (even in high-income countries) and ad hoc recycling or other informal waste collection (especially in low-income countries). In addition, we did not address international import and export of waste." Jambeck et al., "Plastic Waste Inputs," 770. The assumptions built into the parameters of the research design—per capita waste averaged across wealth disparities, population counts that similarly average out differences in waste practices, and the assumption that 50 kilometres of coastline, worldwide, means the same kind of thing—are gross estimations. It is simply impossible math, pure charisma. While the report does say that its goal is to calculate plastic waste numbers in terms of orders of magnitude rather than with any kind of precision or accuracy, it also produced a map that assigns responsibility for global plastic pollution (below I will also note that the estimates have proven incorrect by orders of magnitude). The math accomplishes the universalizing of difference, context, and history to a shockingly high level, even for science. The lead author on the paper, Jenna Jambeck (unmarked), has noted on her website, "We had to use country-level data to build out our framework—so we do indeed have a list of countries that are top contributors. And this has been getting a lot of attention so I want to be clear about how we think about this list—it is not about finger pointing." Jambeck, "Plastic Waste Inputs from Land." Yet this is almost exclusively how the paper has been used, not least because it accurately reflects how the paper is framed. This is another example of how colonialism is accomplished through science and statistics, not shitty intentions.

For more on colonialism and statistics, including resistance against colonial concepts and premises that are built into data, see Walter and Andersen, *Indigenous Statistics*. Also see anything by the excellent Desi Rodriguez-Lonebear, a leading force and a pathmaker for the next generation of Indigenous data scientists. E.g., Rodriguez-Lonebear, "Building a Data Revolution in Indian Country." Maarsi, Desi, for your work, your way of working (Ethics! Commitments! Humour! Brilliance!), and your collegiality. You are my hero.

127 Mai et al., "Riverine Microplastic Pollution in the Pearl River Delta, China." Modeled plastic outflow based on "mismanaged waste" was 91,000–170,000 tonnes per year, while the measured outflow was 2,400–3,800 tonnes per year.

There is a long history of judging a society's waste management as a proxy for its level of civilization. This discourse is fueled by dominant Western frameworks where "human beings, both at an early age individually, and in societies at 'less developed' phases of Civilisation, are profoundly coprophiliac. They love the sight and smell of their own wastes, or at any rate are not disgusted by them. But as Civilisation historically develops, these initial coprophiliac dispositions are brought under increasingly rigorous control, just as the child in the contemporary West is toilet-trained out of its initial lack of revulsion towards its own faeces."[128] In short, from a colonial point of view, models of waste management are tied to ideas about civilization (European self-portraiture) and morality.

The "rigorous control" of plastic wastes from a colonial perspective includes practices such as municipal curbside collection of trash and recyclables, industrial-scale recycling, and highly controlled and technical landfilling—the cultivation of a containment system based on assumed ontologies of separation. Historically, these models originate in US systems and standards. The sanitary landfill was invented in Fresno, California, in 1935, around the same time that Streeter and Phelps were trying to manage sinks on the Ohio River.[129] Practices of waste management that fall outside of this careful cultivation by containment are categorized as "mismanaged" by Jambeck and colleagues. Whether the focus is on agriculture or waste practices, "those who maintained subsistence modes of cultivation, for instance, [are] cast as in need of improvement through assimilation into a civilized (read English) population and ways of living."[130] They are in deficit and need to learn to manage better, or be managed.[131]

128 Inglis, "Dirt and Denigration," 212. This evolutionary model is central to some of Sigmund Freud's theories of development. Also see Chakrabarty, "Open Space/Public Place"; Desai, McFarlane, and Graham, "Politics of Open Defecation"; and Doron and Raja, "Cultural Politics of Shit."

129 Though metal and other scrap recycling is centuries old, curbside recycling of disposables began in the northeast United States in the 1960s and 1970s, with sponsorship from the container and beverage industries, to allow the continual outsourcing of waste disposal. The 1970s also saw the invention of the plastic beverage bottle (made of PET plastic) by DuPont, an American-based chemical company. Weird. Except it's not weird, as shown in Hawkins, "Performativity of Food Packaging"; Elmore, *Citizen Coke*; and Strasser, *Waste and Want*. For more context on the intersections of water and waste via plastic, see Pacheco-Vega, "(Re)Theorizing the Politics of Bottled Water," 658.

130 Bhandar, *Colonial Lives of Property*, 36.

131 The English word *cultivation* "comes from the Latin verb *colere*—to till, tend, or care for—cultivation is a term for social practice rooted in engagements with nonhuman nature." I bring this up because it is an example of the structural violence that care can accomplish. Moore, Kosek, and Pandian, introduction, 9.

The Jambeck study is cited heavily in the Ocean Conservancy's 2015 report, "Stemming the Tide: Land-Based Strategies for a Plastic-Free Ocean."[132] The report's key proposal is to burn 80 percent of the waste in coastal Asia—those countries listed in the Jambeck study—to mitigate marine plastic pollution. To help this happen, they advocate changing national laws to allow foreign companies to build incineration infrastructure in the region. This is the new-again face of waste colonialism.[133] Again, people do not have to be jerks to maintain and reproduce colonial relations—they can have benevolent environmental goals. They can be working to solve important scientific questions. If land relations are colonial, the solutions, initiatives, and studies that flow from those relations will also be colonial.

Producing Difference

The first time I presented the concept of pollution as colonialism, it was early 2018 and the Stanley verdict had just come down, like a ton of bricks, two days earlier. Saskatchewan farmer Gerald Stanley (settler) had been acquitted of any wrongdoing for shooting twenty-two-year-old Colten Boushie (Cree) in the back of the head for trespassing on private land in 2016. As I presented theories of colonialism, I felt sick to my stomach with rage, grief, repulsion, and helplessness. The ability to end the life of Indigenous people, Black people, and people of colour[134] for being on Land that is currently considered private property (or even common property such as sidewalks and city streets) follows similar logics as the right to pollute.[135] Pollution is about maintaining differentiation through

132 For writing against this kind of colonial management of plastic wastes, see GAIA Coalition, "Open Letter to Ocean Conservancy."

133 M. Liboiron, "Waste Colonialism." Also see Reed, "Toxic Colonialism"; Kone, "Pollution in Africa"; and Pratt, "Decreasing Dirty Dumping."

134 I know there are discussions about *BIPOC* as a collective term, and they are good discussions. I continue to use it because I think it does something similar to what LGBTQ2SAI+ does; it shows that there is diversity within a genre of people and we all know that some of the *Q*s are also 2*S*s and maybe *T*s, and that doesn't break the acronym. Also, historically these acronyms will keep shifting and evolving. When I was a kid there was only *L* and *G*. Now look at us! I anticipate a similar future for *BIPOC*.

135 Trying to remember when I first learned this lesson, and whom I might cite, I recall a children's song we would sing on the middle-school playground. The tune is set to settler Woody Guthrie's (unmarked) colonial anthem, "This Land Is My Land":
This land is my land,
This land ain't your land.

appropriation and access to land, about keeping it reserved for settler goals and unavailable for other Land relations. So, too, is shooting unarmed Black, Indigenous, and people of colour.

Activist-geographer Laura Pulido (Chicanix) writes, "Land is thoroughly saturated with racism. There are at least two primary land processes to consider: appropriation and access. Appropriation refers to the diverse ways that land was taken from native people, as previously mentioned. Once land was severed from native peoples and commodified, the question of access arose, which is deeply racialized. Numerous laws and practices reserved land ownership for whites."[136] Some of these practices, like Stanley's, include guns. Others use pipelines. Still others measure the velocity of the Ohio River. They all guarantee colonial and settler access to Land for colonial and settler goals. Aileen Moreton-Robinson (Geonpul, Quandamooka First Nation) contends that property is best understood as a measure of what the white settler is capable of claiming as their own.[137] Pollution and wasting (of Land, of life) does not just accrue value and right to access to whites and settlers. It *produces* whiteness and settlement. As law theorist Cheryl Harris (Black) has argued, "The right to use and enjoyment, the reputational value, the power to exclude, are all characteristics of whiteness shared by various forms of property. Whiteness is . . . an analogue of property,"[138] as it is a condition and producer of white subjectivities—proper, civilized citizens. For all people, "Our ontological relationship to land is a condition of our embodied subjectivity,"[139] whether that relationship is through private property, Resource, kin, or Land.

I got a shotgun,
And you ain't got one.
If you don't get off,
I'll blow your head off.
This land is private property.

136 Pulido, "Geographies of Race and Ethnicity II," 529.

137 Moreton-Robinson, "Problematics of Identity." Thank you, Aileen Moreton-Robinson, for your work. In addition to *The White Possessive*, your presentation at the Native American and Indigenous Studies Association (NAISA) about how Indigenous identity is also able to be possessed (via appropriation and access!) by settlers and whites, and your call to move politics beyond identity, has resonated for me personally, professionally, intellectually, and politically. Thank you.

138 Bhandar paraphrasing Harris in Bhandar, *Colonial Lives of Property*, 7, from Cheryl I. Harris, "Whiteness as Property."

139 Moreton-Robinson, *White Possessive*, 17.

I remember a Q&A session after Michelle Murphy (Métis) presented on colonialism and pollution at the Society for Social Studies of Science (4S), where she asserted that even breathing had become a threat to Black people, Indigenous people, and people of colour.[140] A white woman raised her hand and countered Murphy's arguments by saying she didn't feel threatened by breathing or pollution, even though both are ubiquitous. Rather than arguing against the point, the questioner proved it: breathing is only dangerous to some.[141]

Unevenness is a defining feature of pollution:

> Toxicity is produced by and reproductive of different orders of life. Here, we articulate harm as that which disrupts order and existing relations, while also showing that toxic harm also maintains systems, including those that produce inequity and sacrifice. Then, we turn to toxic politics—struggles pertaining to power focused on which forms of life are strained or extinguished while other forms reproduce and flourish. . . . More than just the contravention of an established order within a system, toxic harm can be understood *as the contravention of order at one scale and the reproduction of order at another*. [For example,] chronic low levels of arsenic in water interrupt the reproduction of fish but maintain the ability of mining companies to store mining tailings in open air mounds.[142]

Unevenness is not only a description of pollution and its harms, not only a side effect, and not remotely an accident: unevenness is an accomplishment of pollution. It is its goal.

The Basics

I began this chapter by saying we would start with the basics: Land, Nature, Resource, and Property. I've attempted to denaturalize how these terms are usually used so the land relations that allow them to make sense become apparent. I hope I've explained it so that when you're on a call with the federal government and someone says that we should set a quantifiable limit to plastics in an environment, you can know this is about enclosure, access, private property, and whiteness. You might choose to tell just a small part of the story—perhaps

140 Murphy, "What Can't a Body Do?"

141 E.g., T. Allen et al., "I Can't Breathe"; Dillon and Sze, "Police Powers and Particulate Matters"; Simmons, "Settler Atmospherics—Cultural Anthropology"; Pulido, "Geographies of Race and Ethnicity II"; and Choy, "Air's Substantiations."

142 M. Liboiron, Tironi, and Calvillo, "Toxic Politics," 333, 335; emphasis in original.

about universalism, or maybe about maximum use of sinks—so the scientists will understand without having to sit them down for days and days of stories. If you are reading this as an academic, expert, teacher, and/or storyteller, I think this is an important part of our jobs—discerning when to tell the whole story, when to tell parts, which parts to tell for which audiences, knowing that they interlock, but being able to pull the parts apart when needed.[143]

This chapter is about modern environmental pollution, but it has also been crafted as an invitation for you to look at the structuring logic of your own discipline and forms of knowledge creation to see what its land relations are, what might be colonial about it, and which naturalized and seemingly benign techniques grant access, moralize maximum use, universalize, separate, produce property, produce difference, maintain whiteness. If our methodological interventions do not address land relations, then they don't address colonialism—we just end up with another study on "mismanaged" waste and another Stanley verdict. Let's do better.

143 This technique is shown beautifully in Dimaline, *Marrow Thieves*.

2 · Scale, Harm, Violence, Land

Not all pollution is colonial. This isn't a statement about its origins in imperialist hands or who its downstream recipients are. Pollution can be colonial without affecting Indigenous people. As la paperson argues, "Primitive accumulation involves not only the gathering of 'natural' resources as assets but also externalizing the 'cost' of the accumulation in the form of contaminated water, disease, and other traumas to the 'natural,' nonpropertied, that is 'indigenous,' world. To be subject to anti-Indian technologies does not require you to be an Indigenous person."[1] There are and can be other ways of thinking about and enacting pollution (or not enacting it) that do not assume access to Indigenous life, land, and bodies. This chapter will point to some of the already existing articulations of pollution-otherwise.

What Is a Chemical?

In a basement room at the University of Waikato, Aotearoa, feminist thinker Michelle Murphy (Métis) asks a room full of (mostly) Indigenous scholars, "What is a chemical?"[2] Murphy had just spoken about the aging Imperial Oil refinery in Sarnia's Chemical Valley on Aamjiwnaang First Nation land, link-

1 paperson, *A Third University Is Possible*, 11.

2 Murphy, "Data towards Dismantlement." Also see Shadaan and Murphy, "Endocrine-Disrupting Chemicals (EDCs) as Industrial and Settler Colonial Structures," which builds on this talk.

ing financialization,[3] infrastructure, and data reporting regulations. Murphy painted a picture not only of the power structures that maintain the petrochemical industry, but also of how Indigenous uses of science and technology might leverage industry-produced data for anticolonial ends. A few things were not mentioned: the scientific names, molecular structures, or individual and epidemiological effects of the chemicals in question. Instead, Murphy was showing that both pollution and its industrial chemicals are best understood not as wayward molecules, but rather as regimes of living, ways of living with and within colonial political economies—what I have been calling Land relations. Murphy writes, "Chemical regimes of living, in which molecular relations extend outside of the organic realm and create interconnections with landscapes, production, and consumption, [require] us to tie the history of technoscience with political economy."[4] In short, industrial pollutants like carbon dioxide or marine plastics are not discrete actors (bad or otherwise) but a set of relations.

Elizabeth Hoover (Mohawk/Mi'kmaq) describes these relations in her book *The River Is in Us*. Hoover tells the story of how Akwesasne community members navigate the contamination of their land in a way that shows the inextricable ties between health, identity, day-to-day practices, tradition, family, and the environment, denaturalizing biomedical concepts of health and harm. Hoover's work moves toward understanding "a shared history of sovereignty, capitalist encounters, resistance, and integrated innovation."[5] Taking my cue from Murphy's and Hoover's models of chemicals as complex relations rather than discrete and autonomous entities, this chapter looks at how chemical relations become apparent even in pollution science and toxicology, the forces that also invent chemicals as discrete bad actors.[6]

As a science and technologies scholar, I understand chemicals as models and metaphors that represent a particular understanding of the world rather than

3 Murphy's presentation built on the work of Chief Arthur Manuel (Secwepemc). See Manuel and Derrickson, *Unsettling Canada*. Also see Pasternak, "Transfer Payments"; and Pasternak, "Mercenary Colonialism."

4 Murphy, "Chemical Regimes of Living," 697.

5 Hoover, *River Is in Us*, 249, quoting McMullin, *Healthy Ancestor*, 159.

6 Murphy and Hoover are just two scholars who describe chemicals as relations rather than wayward and misbehaving molecules. Others include Simmons, "Settler Atmospherics—Cultural Anthropology"; Agard-Jones, "Bodies in the System"; Balayannis and Garnett, "Chemical Kinship"; Masco, "Side Effect"; Cram, "Becoming Jane"; Altman, "American Petro-Topia"; Wylie, *Fractivism*; Shapiro, "Attuning to the Chemosphere"; Landecker, "Food as Exposure"; and Carson, *Silent Spring*. Also see the special issue of *Catalyst* (2020) on chemical kinship edited by Angeliki Balayannis and Emma Garnett.

a faithful mirror of it, "a kind of rendering—a partial representation that either abstracts from, or translates into another form, the real nature of the system or theory, or one that is capable of embodying only a portion of a system."[7] The way dominant science[8] understands chemicals often operates from a colonial worldview that privileges separation and discreteness within Nature. In this worldview, the expert understands chemicals from a god-like,[9] above-it-all, and looking-from-the-objective-outside[10] scientific position. As a scientist, I also understand plastics as polymers characterized by long, strong molecular bonds that allow extreme longevity, even while they fragment down to the nanoscale. These polymer molecules allow monomers (shorter molecular chains) like bisphenol A (BPA), polychlorinated biphenyls (PCBs), and other endocrine-disrupting compounds (EDCs) to nest among their structure. Such a loose arrangement allows monomers to leave their polymer hosts and end up in landscapes and bodies, where their shapes mimic hormones and cause harm. I also understand them the way Murphy and Hoover do, as complex relations, as Land. These understandings are not mutually exclusive or even at odds.

This chapter weaves these two understandings together to move beyond the usual articulation of plastics and their chemicals as autonomous, wayward particles that cause harm to instead talk about scales of colonial violence. This is not an academic exercise. It is the groundwork we need to do anticolonial science, research, and activism in ways that decrease the reproduction of colonial land relations by positing other types of relations with plastics.

After articulating a theory of scale to help distinguish between harm, violence, and the different justices that address them, I will tell a story of bisphenol A (BPA), a common industrial chemical in polycarbonate and other plastics, to show how both colonial and anticolonial moments exist in dominant science. While dominant science has the potential to operate in ways that understand Murphy's call to redefine chemicals, I show how most mainstream activism around plastic pollution does not. I end the chapter with examples of living with pollution in ways that are not generally anticipated by either dominant science or environmental activism, but that instead come out of anticolonial land relations.

7 Morgan, Morrison, and Skinner, *Models as Mediators*, 27. For how these renderings can include intimacy and care, see N. Myers, *Rendering Life Molecular*.

8 See footnote 77 in the introduction on why I use the term *dominant science* instead of *Western science*.

9 Haraway, "Situated Knowledges."

10 Daston, "Objectivity versus Truth."

Scales of Harm and Violence

Scale is not about relative size. Scale is about what relationships matter within a particular context.[11] For example, if you look at live skin cells under a microscope, you'd notice osmosis (flowy water movement) and other activities taking place within the cellular membrane. If you put away the microscope and look at the arm the skin cells come from, you'd notice goose bumps and tattoos. Even though arms are made of skin cells, they do not act like skin cells, and chopping an arm into little pieces does not produce skin cells—though it might produce a prison sentence. Cells and arms are fundamentally different things, even though they are intimately, inextricably related. Scale is a way to talk about this ontological shift, where the processes that matter (relaaaatioooooons!) are of a fundamentally different sort at different scales without severing relations that cross scales. My example from the introduction of the difference in your relations with your daughter versus the mail carrier is an example of scale: who is closer? What kinds of relationships and obligations matter, depending on how "close" someone is? You can talk about the differences between types of kin as scale.

One of the things I do as a scientist is describe the number of plastics ingested by fish, birds, and seals in human food webs. Atlantic cod around the island of Newfoundland have a plastic ingestion rate of 1.68 percent, meaning if you catch a hundred cod, two-ish of them will have ingested plastic.[12] That's my job. I get animal guts[13] from hunters and fishers and look inside for plastics. One of the fish CLEAR has studied is silver hake, a white fish you might know best as the fake crab in sushi. With changing water temperatures, silver hake is starting to make its way up to the island of Newfoundland. But something weird happened: we looked in 134 silver hake guts and found zero plastics.

At first, I thought my students had made a mistake. I checked their work and it was good. We went to the literature that rarely (read: never) reported 0 percent ingestion rates to see if there were any other species that didn't ingest plastics to

11 These ideas on scale (and many more!) were developed with Erica Robles-Anderson (unmarked) during a postdoc with Intel's Science and Technology Center in Social Computing (ISTC-S) in 2012 and beyond. Thank you, Erica, for the intellectually and emotionally generous space you created during our work together. Rarely has it been so bloody exciting and joyful to think together with someone for so long. Though we never finished our scale paper, I value our collaboration immensely. Thank you so very much for this quality of intellectual discussion and setting the bar so high.

12 M. Liboiron et al., "Low Incidence of Plastic Ingestion among Three Fish Species."

13 *Guts* is a term of art that refers to the entire gastrointestinal tract of an animal, from mouth to anus. A fish has an anus, but a bird has a cloaca. *Guts* has them both covered.

see if we could explain the hake anomaly. We found something weird. On closer inspection, most scientific studies compiled multiple species together from a single region and provided an overall plastic ingestion rate. When we disentangled published data so that species were sorted out individually, we found that 41 percent of *all* fish species reported in the scientific literature did not ingest plastics. The 0 percent ingestion rates were being obfuscated by these regional groupings.

Excited, we published a paper proudly proclaiming our finding now that we knew a 0 percent ingestion rate was common.[14] And then the hate mail poured in, accusing me[15] of working for the plastics industry and doing sketchy science, of having said that plastics don't cause harm.

The problem with assuming that a 0 percent plastic ingestion rate is the same as saying that plastics are all right is what I call a scalar mismatch. It conflates relationships, specifically the foraging habits of one species of fish with the violence of a system that allows plastics to exist in every environment ever tested. The ubiquity of industrially produced plastics makes them bioavailable (available to be eaten) to the animals that live in nearly any environment, regardless of whether those animals also eat the plastics. In short, instead of focusing on harm (the effects of plastics on a particular species of fish) we can look at violence, which is the origin of potential harms.[16] Regardless of whether I find plastics in any given fish species, the pipeline that moves plastics into waterways remains the same. We can move from a question of harm that asks "how much" (a question based on threshold theories) to "how" and "why" questions about violence (the relational questions that matter at a different scale).

14 F. Liboiron et al., "Zero Percent Plastic Ingestion Rate." By the way, we don't know why silver hake don't eat plastic. That's a question for laboratory fish scientists. We do know that some animals, like zebra fish, can tell plastics are inedible and spit them back out, and we know that some foraging and feeding habits make some species of animal more susceptible to ingesting plastics than others, but we really can't explain exactly why Atlantic cod have an ingestion rate of 1.68 percent and hake have 0 percent and another species has 23 percent. Kim et al., "Zebrafish Can Recognize Microplastics as Inedible Materials."

15 Even though I am not the first author on this paper (I'm last author, the place that lab granddaddies go), I was the only one accused of sketchy science. I suspect this is because I'm the one that does the lab's social media, so I was the one posting the results online. Plus, France (the first author) and I have the same last name. We've asked our nans to look into it and we are somehow not related, even though we're the only Liboirons we've each met that we didn't already know. We're probably related but forgot.

16 This argument, in snapshot form, was published in *The Conversation* as a public piece following my hate mail. I don't know how much it helped, though it's had a lot more views than the academic paper! M. Liboiron, "Not All Marine Fish Eat Plastics."

Leaving silver hake for a moment, let's talk about BPA, a chemical found in hard plastics like water bottles and the shiny coating on cash register receipts. BPA has been linked to recurrent miscarriages, early-onset puberty, early-onset menopause, obesity, diabetes, cancer, decreases in reproductive health, and neurological disorders like early-onset senility in adults and reduced brain development in children.[17] This list is all about harm. Let's scale up to violence. BPA is one of the highest-volume industrial chemicals produced worldwide, "with over 6 billion pounds produced each year and over 100 tons released into the atmosphere by yearly production."[18] BPA has been found in 91 percent of the tested Canadian population,[19] 93 to 95 percent of the tested US population,[20] 100 percent of tested fetal cord serum in California,[21] 98 percent of the tested elderly population in Sweden,[22] 97 percent of the tested Chinese population,[23] 98 percent of the tested Belgian population[24]—you get the statistical drift. BPA is water soluble, meaning that the body metabolizes it and can excrete its metabolites (pee it out) within six hours.[25] Given how quickly it cycles out of the body, exposure to BPA must be ubiquitous for it to turn up in biomonitoring tests at such high rates. A study that eliminated common sources of BPA in a family's diet reduced their BPA by 66 percent, with the other 34 percent hypothesized to come from outside sources. The authors of the study noted that, even with control of the subjects' diets, "it is difficult to determine exactly which of these changes in food sourcing and handling were responsible for the significant [66 percent] exposure reductions we observed."[26]

The ubiquity of BPA means its relations are operating on a scale different

17 Bergman et al., "Impact of Endocrine Disruption," A104; World Health Organization and United Nations Environment Programme, "State of the Science of Endocrine Disrupting Chemicals."

18 Vandenberg et al., "Bisphenol-A and the Great Divide."

19 Bushnik et al., "Lead and Bisphenol A Concentrations in the Canadian Population."

20 Calafat et al., "Urinary Concentrations of Bisphenol A and 4-Nonylphenol"; Calafat et al., "Exposure of the US Population to Bisphenol A and 4-Tertiary-Octylphenol."

21 Gerona et al., "Bisphenol-A (BPA), BPA Glucuronide, and BPA Sulfate in Midgestation Umbilical Cord Serum."

22 Olsén et al., "Circulating Levels of Bisphenol A (BPA) and Phthalates in an Elderly Population in Sweden."

23 He et al., "Bisphenol A Levels in Blood and Urine in a Chinese Population."

24 Pirard et al., "Urinary Levels of Bisphenol A, Triclosan and 4-Nonylphenol in a General Belgian Population."

25 Bushnik et al., "Lead and Bisphenol A Concentrations in the Canadian Population."

26 Rudel et al., "Food Packaging and Bisphenol A and Bis(2-Ethyhexyl) Phthalate Exposure."

from those that matter for harm. The BPA present in 91 percent of Canadians doesn't guarantee biomedical harm to Canadians. Instead, its ubiquity is better understood as violence, a manifestation of the permission-to-pollute system that allows BPA to be found in nearly all those tested. You don't need to prove you're sick from BPA to make the argument that the ubiquitous contamination of people by an industrial chemical is fucked up.

Instead of defining violence as a direct event of force or coercion, the concept of structural violence "directly illustrates a power system wherein social structures or institutions cause harm to people in a way that results in maldevelopment or deprivation . . . that constrain[s] them from achieving the quality of life that would have otherwise been possible."[27] Structural violence affects different people differently, creating and solidifying social differences and stratification, the basis for reproductive injustice.

The Native Youth Sexual Health Network (NYSHN) and its allies have been using and developing the term *environmental violence* to describe this phenomenon for nearly a decade. The NYSHN defines environmental violence as

> the disproportionate and often devastating impacts that the conscious and deliberate proliferation of environmental toxins [*sic*][28] and industrial development (including extraction, production, export and release) have on Indigenous women, children and future generations, without regard from States or corporations for their severe and ongoing harm. Furthermore, since 2010, NYSHN's work around the term has fostered recognition of the ways it has evolved to not only include the biological reproductive impacts of industry, but also the social impacts. This work has been critical in recent years, as attention paid to the threats of industry in Indigenous communities has tended to focus entirely on the biological health impacts of fracking and mining, or entirely on the sexual violence acts stemming from the male population booms of industry workers' camps. Rarely is attention paid to both types of impacts, with recognition of their intimate connection to the land.[29]

27 B. Lee, "Causes and Cures VII."

28 Scientifically speaking (which is not the only or even best way to speak, of course), toxins are poisons produced in animal cells and other nonindustrial sources. Toxicants are industrial chemicals produced in labs. I've written about why it's important not to conflate toxins with toxicants (spoiler: it's about scale): M. Liboiron, "Toxins or Toxicants?"

29 For the record, I think this report is one of the best I've read that understands chemicals as relations rather than objects. Native Youth Sexual Health Network and Women's Earth Alliance, "Violence on the Land, Violence on Our Bodies."

Kyle Powys Whyte (Potawatomi) argues that climate change and other forms of using Land as a sink "systematically erase certain socioecological contexts, or horizons, that are vital for members of another society to experience themselves in the world as having responsibilities to other humans, nonhumans and the environment."[30] Environmental violence is about who gets to erase—or produce—and how that is structured so that pollution becomes normal, even ubiquitous.

Scientific methods for monitoring marine plastic pollution have the capacity to "study up" toward structures of violence. Instead of conducting baseline studies that focus on the weight, polymer, and type of plastics found in marine life, the #breakfreefromplastic movement conducts brand audits in which contributors record the brand identities that appear on plastics found on shorelines.[31] The result is a tally of which companies produce the plastics in local environments. Von Hernandez (Filipino), #breakfreefromplastic's global coordinator, says, "Corporations cannot greenwash their role out of the plastic pollution crisis and put the blame on people, all the time. Our brand audits make it clear which companies are primarily responsible for the proliferation of throwaway plastic that's defiling nature and killing our oceans. Their brands provide undeniable evidence of this truth."[32] On Freedom Island, Philippines, "six international brands are responsible for 53.8% of plastic packaging pollution."[33] These brands are Nestlé, Unilever, PT Torabika Mayora, Procter and Gamble, Monde Nissin, and Colgate Palmolive. To my knowledge, the findings of #breakfreefromplastic's brand audits have not been replicated in any other scientific study. Time to start.

Describing structural violence allows interventions to occur on the right scale to impact salient relationships. Though methodological discussions of "studying up" to investigate sources of violence, rather than focusing on harm

30 Whyte, "Indigenous Experience, Environmental Justice and Settler Colonialism."

31 By the time plastics have broken down into microplastics (less than 5 millimetres [mm]), brand identification is usually impossible. But when plastics are larger than 5 mm, they often contain some brand identity such as labels, logos, and bar codes. Larger plastics also tend to be more local plastics, as the longer that plastics are in the water or on shorelines, the more likely they are to have fragmented into microplastics. Brand audits are an excellent example of accountability metrics in pollution research. GAIA, "Plastics Exposed."

32 Mundo, "#breakfreefromplastic Is Supercharging Coastal Cleanups."

33 #breakfreefromplastic, "Green Groups Reveal Top Plastic Polluters." Similar brands also pollute other areas. The brand audits place the data for all their studies in an online repository. You can see current numbers for different areas at http://plasticpolluters.breakfreefromplastic.org/.

and victimhood, occur almost exclusively in the social sciences, brand audits are one example of how the methodology is not discipline-specific and can take place in the natural sciences.

Plastic Relations

The remainder of this chapter investigates scales of harm and violence in what Michelle Murphy calls *alterlife*, or "the condition of being already co-constituted by material entanglements with water, chemicals, soil, atmospheres, microbes, and built environments, and also the condition of being open to ongoing becoming. Hence, alterlife is already recompiled, pained, and damaged, but has potentiality nonetheless. If life holds together tensions between violence and possibility, braiding the organic and inorganic, body and land, and resides in the indistinctions between infrastructures and ecologies, recognizing alterlife attends also to openness, to a potential for recomposition that exceeds the ongoing aftermaths."[34]

What kinds of science and activism are suited to an alterlife characterized by ubiquitous and permanent pollution by plastics and their chemicals? How can these scientific and activist actions address violence, rather than only harm? At what scales do plastic relations work and how do we address them at the same scale? Working on these questions will give us some knowledge about the relationships that matter (scale) and lay the foundation for our anticolonial work. The good news is that there is already considerable activity that exemplifies alterlife living with plastic pollution.

The Strength of the Curve

In 1982, some lab rats posed a problem. Scientists within the US Department of Health and Human Services' National Toxicology Program carried out "a series of experiments designed to determine whether selected chemicals produce cancer in animals."[35] The series included a study on BPA. Rats were fed BPA at levels between 1,000 and 10,000 parts per million (PPM), or 1,000 to 10,000 milligrams of BPA in one litre of water. Scientists found "no convincing evidence that bisphenol A was carcinogenic for F344 rats or B6C3F1 mice of either sex" but found that all treated rats exhibited reduced body weight compared to controls.[36]

34 Murphy, "Against Population, towards Alterlife," 118.

35 National Toxicology Program, "Carcinogenesis Bioassay of Bisphenol A," ii.

36 National Toxicology Program, "Carcinogenesis Bioassay of Bisphenol A," ii. Though the report stated there was "no convincing evidence" of carcinogenic activity from BPA, it also cautioned that this statement should be qualified to reflect the facts that leukemia in

The US Environmental Protection Agency (EPA), the beneficiary of the report, believed that "reduced body weights in rats . . . [were] considered a direct adverse effect of exposure to bisphenol A." They used the study to determine a regulatory oral reference dose (RfD) for the chemical.[37] Oral reference doses measure assimilative capacities of bodies.[38] They are premised on the idea that thresholds of safety exist for certain toxic effects below which harm is not scientifically observed in cells, organs, organisms, or populations.[39]

Except there was no evidence of threshold in the study. At the study's lowest dose (50 milligrams [mg] / kilograms [kg] of body weight/day), rats exhibited reduced body weight. The EPA noted that the study "failed to identify a chronic NOAEL [No Observable Adverse Effect Level, or threshold] for reduced body weight."[40] NOAELs are one type of threshold measure based on assimilative capacity; it's the moment before adverse effects at the cellular or organ level are observed, and like Streeter and Phelps's oxygen sag, NOAELs are represented as sigmoid curves. The EPA used the study data to predict where the threshold would have occurred had lower doses been used, depending on the universality of the sigmoid curve that they assumed all toxic relationships exhibit. Based on this extrapolation, they set the maximum oral RfD for humans in the United States at 50 micrograms [μg] / kg of body weight/day.[41] This threshold is still in

---

male rats showed a significant positive trend, that leukemia incidence in high-dose male rats was considered not significant only on the basis of the Bonferroni criteria, that leukemia incidence was also elevated in female rats and male mice, and that the significance of interstitial-cell tumors of the testes in rats was dismissed on the basis of historical control data (ix). This study was fucked up. Sara Vogel (unmarked) has done an excellent job detailing the ethical and methodological problems with the study, including fraudulent activity and plain old bad research design. These BPA studies have been reviewed more recently by J. Huff (unmarked) because, using "the approach to interpreting cancer in animal studies used today, BPA would be interpreted as being associated with an increase in tumors of blood cells, the testes, and the mammary glands" (281). But the point of this story is not that the study was Bad Science or Old Science, but that its conclusions made enough sense to be taken up in regulation, regardless of the goodness, badness, or outdatedness of the laboratory methods. See Vogel, "Politics of Plastics"; Huff, "Carcinogenicity of Bisphenol A Revisited"; and J. Myers et al., "Why Public Health Agencies Cannot Depend."

37 US Environmental Protection Agency, "Bisphenol A."

38 See chapter 1 for the full story on assimilative capacity.

39 For thorough research on the logistics and politics of constructing a regulatory safe dose, see Vogel, *Is It Safe?*

40 US Environmental Protection Agency, "Bisphenol A."

41 The EPA was confident "that the NOAEL for reduced body weight in rats is probably not far below the LOAEL [Lowest Observed Adverse Effect Level] of 1000 PPM of the diet." The

use today. In short, the EPA's confidence in the perfect regularity of toxicological sigmoid curves is the basis of the regulation designed to keep us safe today, *even though there was no empirical evidence for the regulated quantity*.[42]

Challenging the Curve

In another laboratory, this time in 1998 in Cleveland, Ohio, endocrinologist Patricia Hunt (unmarked) was studying genetic reproductive processes in rats, as one does. One day, 40 percent of her control rats—those not receiving test treatments—displayed a chromosomal abnormality in their eggs.[43] Something had contaminated the rats.

Hunt tracked down the source of these anomalies. It was the water bottles. In the laboratory. With the floor cleaner. She and her team "identified damaged caging material as the source of [a chemical] exposure, as [they] were able to recapitulate the meiotic abnormalities by intentionally damaging cages and water bottles" by washing them in the new floor cleaner the janitorial staff had used, achieving identical results to the accidental exposure.[44] When custodial staff used new cleaner on the rat cages, the cleaner had released BPA from the polycarbonate bottles into the rats' water. Using the trace doses that matched the amount leaching from the water bottles, Hunt's lab found effects at their lowest tested dose of 20 nanograms (ng) per gram (g) (20 mg/kg) of body weight/day, *well below the threshold set by the* EPA *in 1982*. Not good.[45]

EPA expressed "high confidence" in its threshold declaration; they expressed only "medium" confidence in the study itself "because this study, although well controlled and performed, failed to identify a chronic NOAEL for reduced body weight." US Environmental Protection Agency, "Bisphenol A."

42 In 1986, the European Commission's Scientific Committee on Food revised its Tolerable Daily Intake (TDI) (their version of an RfD) of BPA to 0.05 mg/kg of body weight/day. This was again lowered to 0.01 mg/kg of body weight/day in the early 2000s (Hunt et al., "Bisphenol A Exposure," 551). As Johanna Olsson (unmarked) has shown in her excellent dissertation, regulation premised on assimilative capacity is about tweaking where thresholds lie, not challenging the threshold paradigm. Olsson, "Setting Limits in Nature and the Metabolism of Knowledge."

43 Hunt et al., "Bisphenol A Exposure." For more on this story, see Landecker, "When the Control Becomes the Experiment."

44 Hunt et al., "Bisphenol A Exposure." Thank you, Patricia Hunt, for not only your science and your dedication to questions of low-dose effects of EDCs, but also for how you do that science, including your work with the media, journalists, young scientists, and public bodies.

45 Though Hunt's story from 1998 is the most well known, there is another study from 1993 that happened the same way. Scientists in Stanford, California, studying yeast also

In toxicological terms, an observable effect in rats at low doses but not at higher doses means that something is wrong with the threshold theory of harm. It is not just that the curve should be steeper, that the threshold is in the wrong place, or that more data is needed. Hunt's study brings into question toxicology's premise that "the danger is in the dose." BPA's toxicity is not *S*-shaped (sigmoid) at all. Other toxicants also defy the *S*-shaped, threshold-producing curve created by Streeter and Phelps. Radiation and carcinogens can cause harm immediately upon contact, with a straight line between dose and harm rather than offering a threshold before harm occurs.[46] But BPA and its family of chemicals make curves that are neither *S*-shaped nor linear.

If you start to dose rats with BPA at trace quantities and slowly increase the doses, you get a *U*-shaped or wiggly graph called a nonmonotonic curve[47] that "until recently, [was] not considered plausible, and thus they were not published, reported, or interpreted as relevant biological phenomena."[48] That is, nonthreshold theories of toxicity were inconceivable. While nonthreshold phenomena like carcinogens and radiation were well plotted and examined, the unique genre of dose-response relationships engendered by chemicals like BPA was not initially legible within toxicology.[49] The dawning recognition that such

---

found that BPA was leaching from polycarbonate flasks and published the results. Yet it was Hunt who pursued the issue in scientific and public forums, making it a matter of concern. The 1993 study is Krishnan et al., "Bisphenol-A."

46 Mind you, threshold-thinking is so strong that even in the case of carcinogens and radiation, policy uses risk analysis that allows for a certain amount of population death (or acceptable loss), creating a threshold of harm in a threshold-less form of toxicity. See, for example, Cram, "Becoming Jane." Note from an editor: "It's wild editing this in the midst of coronavirus." For my editor and others coming to understand how they are part of population dynamics in COVID times and would like to see how this logic of disposability was created and normalized, see Murphy, *Economization of Life*.

47 Nonmonotonic curves, also called biphasic or multiphasic curves, are defined as those where "the slope of the dose-response curve changes sign from positive to negative or vice versa at some point along the range of doses examined." Vandenberg et al., "Hormones and Endocrine-Disrupting Chemicals," 380.

48 Lagarde et al., "Non-Monotonic Dose-Response Relationships and Endocrine Disruptors."

49 For the fellow nerds who want to know *why* the curves are different: as I'll discuss below, BPA is one of many chemicals that act like hormones called endocrine disruptors. The hormone system has feedback loops that mean that low quantities of a hormone (or its mimic) have different, and often greater, effects than a large quantity (hence the pico- to nanomole scale at which Hunt and her colleagues work). High amounts of a hormone signal the body to shut down or reduce the synthesis and acceptance of that hormone,

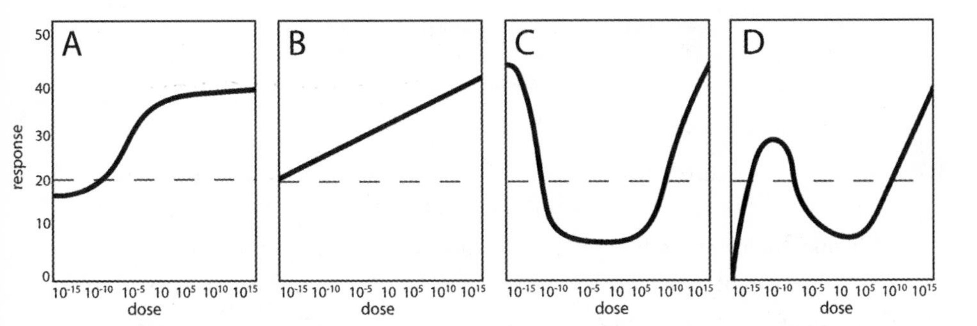

FIGURE 2.1. Different shapes of toxicological curves showing different dose-response relationships. (A) Monotonic, sigmoid threshold curve (e.g., organic pollution). (B) Linear nonthreshold curve (e.g., carcinogens). (C) Simple nonmonotonic curve (e.g., endocrine disruptors). (D) Complex nonmonotonic curve (e.g., endocrine disruptors). Illustration by Max Liboiron. CC-BY 3.0.

---

meaning both hormones and endocrine disruptors often have the greatest effects at the smallest doses. Chemicals that mimic hormones can also have *different* effects at high doses. This is what makes the wiggly, *U*-shaped, nonmonotonic curve rather than the sigmoid curve characteristic of assimilative capacity.

For super-nerds that want even more science (hello! I love you), I quote a review paper on BPA risk assessment at length: "recent findings concerning the multiple mechanisms of action of BPA show that at concentrations < 1 ppt, BPA activates receptors associated with the plasma membrane of selected target cells. As the BPA dose at target increases, various responses in the same or different cells are activated or inhibited, with the specific dose required being dependent on the subtype of nuclear ER and specific coactivators or coinhibitors that are present. At even higher concentrations (parts per billion to parts per million), inhibition of androgen-stimulated and thyroid-hormone–stimulated responses can also occur. That the integrated output across a 1-million-fold dose range can be nonmonotonic (inverted-U shape) is thus not unexpected by scientists who study hormones and hormonally active drugs or chemicals. Regulatory agencies that conduct risk assessments have not addressed the implications of nonmonotonic dose-response curves for endocrine-disrupting chemicals with regard to the linear-threshold model currently." Vom Saal and Hughes, "Extensive New Literature," 931.

relationships do exist has resulted in a paradigm shift within toxicology and related fields.[50]

Toxins versus Toxicants

BPA is not a toxin. It is a toxicant. This might seem like fussy science-nerd semantics, but toxins (animals poisons) and toxicants (industrially produced chemicals) operate at different scales, engender different relationships, have different modes of both harm and violence, and thus have different politics.[51] Toxins, the poisons produced by animals and plants, act at the cellular level. A cell is a thin envelope around a soup of enzymes and organelles. Most poisons work by upsetting the soup. For example, tetrodotoxin, the puffer fish poison popular on criminal forensic TV shows, works by blocking sodium from passing through nerve cell membranes. This breaks down action in cells, which breaks down functions of organs, which causes acute morbidity, then death.[52] Organs can handle some cell death caused by toxins before they stop working, making the *S*-shaped sigmoid curve.

BPA is in a class of toxicants called endocrine disrupting chemicals (EDCs), many of which are found in plastics. EDCs include plasticizers such as BPA and phthalates, flame retardants like PCBs, and pesticides such as DDT and atrazine, among many others. EDCs are industrial toxicants that can participate in the body's endocrine (hormone) system by mimicking hormones. In simplified terms: hormones travel through the body until they encounter a receptor with a shape that complements their own. The hormone and receptor fit together like a lock and key, which signals the DNA in the cell to do whatever it's supposed to do. When an endocrine disruptor shaped similarly to a hormone binds to a receptor, a lot of things can happen: they can get stuck and block hormones from binding and unbinding; they can signal the DNA in the cell to get to work (e.g., expressing genes or making proteins); or they can do nothing at all.

This is a simplistic, biologically reductive way to describe what happens,[53]

50 The following text talks about this shift from within toxicology: Rowlands et al., "FutureTox."

51 I have written about this at length in M. Liboiron, "Toxins or Toxicants?" Often *toxicant* is too froufrou a term when writing for public/diverse audiences, but I don't want to incorrectly use *toxin* in these situations, so I use *toxics* or *toxic chemicals* or something similar.

52 Noguchi, Onuki, and Arakawa, "Tetrodotoxin Poisoning due to Pufferfish and Gastropods."

53 For a more nuanced view, including the role of power and social realms in endocrine relations, see Hannah Landecker's (unmarked) excellent work, including Landecker and Panofsky, "From Social Structure to Gene Regulation, and Back"; and Landecker, "The Social as Signal in the Body of Chromatin."

but the point is that, from the scientific perspective, EDCs do not work like bull-in-a-china-shop toxin trespassers, wrecking things and spilling cell soup. They work as part of the system, disrupting it while allowing it to continue, resulting in things like recurrent miscarriages, early-onset puberty, early-onset menopause, obesity, diabetes, and neurological disorders such as early-onset senility in adults—none of which are directly lethal.[54] This is the type of harm EDCs do, but they are also parts of structures of violence.

I've already mentioned that BPA is one of the highest-volume industrial chemicals produced worldwide, "with over 6 billion pounds produced each year and over 100 tons released into the atmosphere by yearly production."[55] To repeat: BPA is present in nearly all human bodies tested, even though it is water soluble and is peed out in short order. Yet BPA is only one of many, many chemicals in consumer items and the environment. How many? We don't know. Perhaps you've seen the statistic that only 200 of the 84,000 chemicals used commercially have been tested for toxicity (harm to organelles, cells, organisms, etc.), but that number is in an unverifiably large ballpark.[56]

The United Nations Environment Programme's Chemical Outlook report currently estimates the worth of the global chemical industry as US$5 trillion; it is expected to double by 2030, though the program does not provide figures

54 Bergman et al., "Impact of Endocrine Disruption," A104; World Health Organization and United Nations Environment Programme, "State of the Science of Endocrine Disrupting Chemicals."

55 Vandenberg et al., "Bisphenol-A and the Great Divide," 76.

56 The number comes from the United States Toxic Substances Control Act (TSCA) of 1976, which required an inventory of chemicals being produced in the country and resulted in a list of approximately 62,000 chemicals. US Environmental Protection Agency, "TSCA Chemical Substance Inventory." Another 22,000 "new chemicals" were added between 1982 and 2002. That's where the number "84,000 chemicals" come from. But some of these chemicals are duplicates, some are no longer being produced, and some are in the inventory but not on the market (no one knows how many), so the number is smaller. But the number is also much, much larger: TSCA does not cover chemicals involved in food and food additives, drugs, cosmetics, ammunition, pesticides, tobacco, and "mixtures," nor, starting in 1986, do small manufacturers have to report their chemicals if they are produced in quantities under 10,000 pounds. That number increased to 25,000 pounds in 2006. Board on Population Health and Public Health Practice, "The Challenge." With these gaps and compromises, the educated estimate is "there are somewhere between 25,000 and 84,000 chemicals in commerce in the United States." Roundtable on Environmental Health Sciences, Research, and Medicine; Board on Population Health and Public Health Practice; and Institute of Medicine, "The Challenge." This is how much industrial accounting works.

on tonnage per se.[57] We know that global chemical supply chains are growing exponentially. Global production rates for plastics are estimated to increase to about 2,000 million tonnes per year by 2050.[58] Though the data is dispersed and incomplete, one estimate states that half of all plastics made throughout history has been produced between 2004 and 2017, and while this figure is based as much on charisma as math, it does illustrate exponential growth.[59] In terms of toxicity, of chemicals that were tested, an estimated 62 percent of the chemicals consumed in 2016 were categorized as hazardous to health (by volume), impacting at least 1.6 million lives (an acknowledged underestimate).[60] Even when these chemicals are tested for toxicological signs of harm in laboratories, they are rarely if ever tested for their behaviour over time, their degree of chronic exposure, or in contexts where they interact additively with other chemicals (called the "cocktail effect"), meaning that there is little knowledge of how they act in real-life scenarios. Given that industrial chemicals and toxicants are used in an estimated 96 percent of manufactured goods,[61] we are talking about structural issues of exposure that cannot be addressed by focusing on health and harm. The scale and ontology of toxicants, unlike toxins, is about the power to pollute.

In addition to their scales of production, the structural violence of toxicants stems from the dominance of the threshold model of pollution, which "systematically erase[s] certain socioecological contexts, or horizons,"[62] that are vital for kinds of relations that are not industrial. Colonization is not just about having access—it is also about eliminating other types of relations that might threaten that access.[63] Yet some scientists, like Patricia Hunt, have been pushing back against the threshold theory of harm.

57 United Nations Environment Programme, "Global Chemicals Outlook II," vi.
58 PlasticsEurope, "Plastics—the Facts 2016."
59 Geyer, Jambeck, and Law, "Production, Use, and Fate of All Plastics Ever Made."
60 These figures are for Europe only, which does much more testing than Canada or the United States. United Nations Environment Programme, "Global Chemicals Outlook II," vi.
61 American Chemistry Council, "Chemical Production Expanded."
62 Whyte, "Indigenous Experience, Environmental Justice and Settler Colonialism," 3.
63 This brings us back to the tyranny of universalism and theories of Nature based on lawlike behaviours, covered in chapter 1. There cannot be other functional concepts of pollution if Land and bodies are standing reserves for effluents and toxicants.

Scalar Effects: Autonomy and Discreteness Undone

Following Deboleena Roy (diasporic Indian), natural science has its own methods of reflexivity and relationality. She writes, "What if biological reductionism was not seen as an end to scientific knowledge but instead as a means to connect more intimately to multiple microscopic and molecular material actants that make up the world within and around us? What if learning how to *see* the world was also about learning how to *encounter* that world?"[64] In the case of BPA, dominant science has provided its own critique of the hallmarks of colonial science, including autonomy, discreteness, and separation by seeing contaminants differently.

Hunt and other scientists who specialize in hormones have challenged the threshold theory of harm in large part because of the scale of their scientific enquiry—how they work within certain relations. Hunt recounts how she approached dose when her rats were contaminated: "We are geneticists, so we were interested in the lowest levels of the chemical that would affect cellular processes. Most toxicologists would have gone up—on the assumption that, if a little is bad, higher doses should be even worse. Of course, this isn't how hormones work. Hormones work at very low levels, and BPA mimics the hormone estrogen."[65] Hormones have effects at the picomolar to nanomolar range (so small as to be measured by the number of molecules), while toxicology studies happen in the parts per million or billion range, typically measured in milligrams. For reference, if you divided a paperclip[66] weighing 1 gram into one thousand pieces, each would be a milligram. Moles are a chemical numerical unit that works a little differently: one mole is 6.02×10^{23} molecules or atoms of a substance. You need about 56 paperclips to make a mole of paperclip.[67] If you then divided that pile of paperclips into a *trillion* pieces, it would be a picomole. In the first example of milligrams, you're hanging out with paperclip dust. With picomoles, you end up with molecules instead of bits of paperclip. This is a shift in scale, not just relative size, as scientists like Patricia Hunt, Laura Vandenberg[68]

64 Roy, *Molecular Feminisms*, 5–6; emphasis in original.

65 Coombs, "Effects of Exposures on Development of Oocytes."

66 Question from the audience: does a paperclip really weigh 1 gram? Yes, a paperclip weighs 1 gram on a not-very-precise teaching scale. I don't know how many school lab exams I did showing that I knew how to use a scale by weighing a paperclip. Once I got a cork to weigh instead, and I panicked since I didn't already know the answer.

67 This assumes paperclips are made of iron.

68 Vandenberg et al., "Biomonitoring Studies Should Be Used by Regulatory Agencies"; Vandenberg et al., "Bisphenol-A and the Great Divide"; Vandenberg et al., "Hormones and

(unmarked), and many others have found that BPA causes harm at picomole "doses below those used in traditional toxicological studies"[69] that use milligrams. These low doses[70] are "within the range of typical human exposure."[71]

One of the things that allowed scientists across disciplines at the turn of the twentieth century to produce sigmoid curves for the threshold model of pollution was a shift in how they understood disease. In contrast to the miasma theory of disease, which held that bad or smelly air negatively influenced bodies to get sick, new models treated chemicals, germs, and other causes of harm as discrete, with the capability of autonomous action, allowing a scientist to track their influence independently of other relations.[72] As environmental historian Linda Nash (unmarked) has written, "In toxicology research the environment was reduced to a set of discrete chemicals. In fact, the environment of the laboratory is carefully constructed so that agency can be ascribed solely to the chemical under study. Other factors are purposefully eliminated. As bacteriology had collapsed the agency of nature into the agency of a specific pathogen, so modern toxicology had collapsed it into the agency of a specific chemical."[73] This model of discreteness does not work well for EDCs.

Even if you can isolate a particular hormone within the body's soup of hormones, one hormone—or endocrine disruptor—does many things, so the effects are not discrete, even if the agents are. For example, in concert with other hormones, estrogens maintain memory functions, influence fat stores, support lung and heart function, promote mental health, and influence the regulation

---

Endocrine-Disrupting Chemicals"; Vandenberg, Luthi, and Quinerly, "Plastic Bodies in a Plastic World."

69 Vom Saal et al., "Chapel Hill Bisphenol A Expert Panel Consensus Statement," 131.

70 In the toxicological literature, "low dose" refers to two different things. The first meaning (and the one used in this quote) refers to "doses below those tested in traditional toxicology assessments," referring directly to existing, standard, and dominant practices in the field of toxicology. In this use of the term, it is not a set quantity, but the realm of milligrams generally. In some studies, "low dose" also refers to a dose that already exists in a population or environment. Often these two low doses align, but sometimes not. Vandenberg et al., "Hormones and Endocrine-Disrupting Chemicals," 379.

71 Vom Saal and Hughes, "Extensive New Literature," 926.

72 I have written about this change from a relational model of disease and chemical harm to a discrete and autonomous model in M. Liboiron, "Plasticizers."

73 Nash, *Inescapable Ecologies*, 142. For a similar argument about the rise of ecology where ecologists sought to trace complex interactions in an ecosystem by breaking these relations into discrete energy pathways, see Robles-Anderson and Liboiron, "Coupling Complexity."

of metabolism, protein synthesis, blood coagulation, salt and water retention, the development of sexual organs, and the sex drives and fertility of all sexes.[74] Locating "the" effect of an estrogen mimic such as BPA is difficult, if not impossible, and is certainly misled. There are about ninety different hormones, and up to eight hundred chemicals are known or suspected to be EDCs.[75] In a body, these chemicals and hormones "may interact additively, multiplicatively, or antagonistically in what is commonly referred to as the 'cocktail effect.'"[76] The cocktail effect is akin to drinking alcohol, ingesting Nyquil, smoking cannabis, and falling down the stairs in quick succession. While some injuries can be easily assigned to the stairs or one of the specific substances, others are impossible to sort out because all are acting at once, and each chemical changes the behaviours of the others.[77]

Paired with the cocktail effect, the ubiquity of EDCs and plastics also trouble models of discreteness. Nash argues, "When investigators spoke of 'exposure pathways,' they implied that such pathways were narrow routes of entry that could be regulated, tracked, or even blocked. The surface of the body was assumed to be well defined, a boundary between the individual and the outside world that was breached only in specific instances, and exposure itself was assumed to be finite and discrete rather than an ongoing process that involved multiple chemicals on trees, in air, in water, in food."[78] Yet many scientific studies have found that exposure to EDCs is chronic and continuous, as mentioned above.[79]

EDC action also defies expectations that chemicals act at a discrete moment in time, usually at the moment of exposure. In the years since Hunt found that BPA leached from her rats' water bottles into their bodies, she has continued to study the chemical. Hunt became particularly interested in how BPA interacts with developmental processes. In 2007, she released a study in which her lab exposed pregnant mice to BPA just as the ovaries in their fetal offspring were developing. When those fetuses grew up, 40 percent of *their* eggs were abnormal.[80] One moment of exposure affected three generations simultaneously even though some of those generations didn't exist yet.

74 L. Nelson and Bulun, "Estrogen Production and Action."
75 World Health Organization and United Nations Environment Programme, "State of the Science of Endocrine Disrupting Chemicals."
76 Meeker, Sathyanarayana, and Swan, "Phthalates and Other Additives in Plastics," 2108.
77 For a social science treatment of the cocktail effect, see E. Martin, *Bipolar Expeditions*.
78 Nash, *Inescapable Ecologies*, 148.
79 Vom Saal and Hughes, "Extensive New Literature," 926.
80 Susiarjo et al., "Bisphenol A Exposure in Utero."

These intergenerational effects can be understood as artifacts of changes in the genetic structure produced by hormone-mimicking chemicals. But this form of knowledge—this type of understanding—is strictly laboratory knowledge and doesn't work outside of that space. Outside of the lab, intergenerational exposure across three generations takes so many routes, with such extreme latency between exposures and effect, that it is impossible to study them in context. Everyday exposures exceed a causal model. There is no straight line between exposure and effect, particularly given the cocktail of chemicals that bodies are exposed to, the latency of their effects, and the multitude of potential end points.

Instead of understanding intergenerational effects in terms of laboratory results, we can understand them as relations. As geneticist Katherine Crocker (Kanza) has written,

> Our history is neither written by nor coded into our DNA, but it is nevertheless scrawled and carved into us like graffiti. Some things fade quickly but other events last longer, or are temporarily obscured only to resurface generations later, powerful beyond what we have been taught to expect.
>
> Biologists used to have a comfortable dogma. We believed that everything about an organism could be found somewhere in its DNA. That is not wrong, but neither is it the whole story. Now I am here in the Natural Sciences Building, having discovered a new function of hormones in crickets. What I—an Indigenous woman and a scientist in defiance of every obstacle—have found is not limited to crickets. Researchers studying humans have found that our hormones, too, transcend generations and genes. What I have found may be new to biologists, but not to the peoples of what settlers call the New World. We have known for hundreds of generations that we carry our histories within us. They are part of who we are.[81]

Crocker articulates an approach that might be described as the opposite of the separated, discrete, and individualized action that characterizes Nature and its sciences.

Crocker's scientific research, along with that of Hunt and many others, eschews colonial models in favour of other models. Hunt's work shows that anticolonial models do not have to be Indigenous, like Crocker's. There are many ways to ensure that thresholds for allowable pollution, the separation of Nature and culture, the reservation of the future for settler goals, the consignment of

81 Crocker, "Híyoge Owísisi Tánga Itá (Cricket Egg Stories)."

non-Europeans to the back room of civilization, and a fetish for management and ontological discreteness are not reproduced. There are many ways to ensure that other ways of understanding chemicals and pollution flourish. For this reason, plastics are an ideal pollutant to upset dominant norms of pollution—their industrial, intergenerational, and ubiquitous relations make a lot of room for understanding and doing things differently.

Scales of Action That Address Plastics Violence

I am going to assume that some readers will be invested in making change in the world, whether that is through environmental activism or within dominant science. For this, we need to be specific in the relationalities we are taking up, decrying, and changing, including L/land relations.

Purity Activism as Scalar Mismatch

One of the most common interventions into plastic pollution by environmental activists and advocacy groups is consumer avoidance: don't buy water bottles with BPA in them; buy this other polycarbonate water bottle instead. Spoiler: manufacturers often replace BPA with a structurally identical chemical called BPS. Recall that it is the structure of EDCs, which are shaped like hormones, that allows them to do their unique toxicological work in bodies. BPS is shaped similarly to BPA and thus acts similarly to BPA.[82] Even if you swap out your plastic bottle for a glass one, you still have BPA coming in from cash register receipts, paper bills, the lining of canned food, and epoxies. Avoidance, based on the concept of the possibility of separating human (body) and (polluted) Nature, is a scalar mismatch[83] where problems and their proposed solutions occur at different scales and do not affect the relationships that matter. Purity relations based in discreteness and separation do not scale for plastics.[84]

The move from harm to violence helps register the scale of plastic pollution. The scales of problems and the scales of proposed interventions must match if they are to bear on the same thing. Avoidance, consumer choice, and technological fixes respond to scales that miss one crucial relationship in plastic pollution: production. If some amount of plastics will always "leak" out of infrastructures and into oceans, total intentional avoidance of BPA would still result in body

82 Rochester and Bolden, "Bisphenol S and F."

83 M. Liboiron, "Solutions to Waste."

84 Shotwell, *Against Purity*. Also see the special issue on purity: *Excursions* 4, no. 2 (2013); and for relationships between power and purity see Douglas, *Purity and Danger*.

burdens of over 30 percent, enough to cause harm.[85] If plastics and their chemicals are found in tap water, beer, the Arctic, and fetuses, then the relationships we should be looking at are not at the end of the pipe, but in how plastics go into the pipe to begin with.[86]

In the introduction, I noted that plastics have been otherwise. You'll see that figure 2.2, showing the increase in plastics production, starts just prior to 1950, even though plastics have been around since the late nineteenth century. There were plastics before there were mass-produced plastics, but those plastics are never, ever featured on any graphs of plastic production.[87] They simply weren't produced in a way that warrants depiction. The ubiquity of plastics was not inevitable.

Though plastic production has been increasing exponentially since the 1950s, there are two wee wiggles in the graph that show when production fell, if only for a moment. These moments of decline point to the types of relationships that matter at the scale of plastic production and, thus, of pollution. The first dip happened in the mid-1970s, when the energy crisis meant that raw feedstocks of plastics were not as readily available as they had been. The second coincides with the 2008 financial crisis.

Recall that scale is a way to talk about which relations matter within a specific context. We can monitor plastics in the environment and on consumer shelves all we want, but plastics only come from one place: industry. If interventions into plastic pollution have no impact on extraction, financialization, or industry's access to capital, then they aren't going to be effective. Interventions that do not address plastics production can maintain and even secure land as standing reserves for plastics (see chapter 1).[88]

Muddying Harm for Violence

Though plastics cause harm, they also host lives. I'm often asked about bacteria or fungi that "eat" plastics, and isn't that a great way to clean up the oceans and landfills? No. If plastic-eating bacteria or fungi actually scaled to the point

85 Rudel et al., "Food Packaging and Bisphenol A and Bis(2-Ethyhexyl) Phthalate Exposure."

86 Kosuth, Mason, and Wattenberg, "Anthropogenic Contamination of Tap Water, Beer, and Sea Salt"; PAME, "Desktop Study on Marine Litter"; Aris, "Estimation of Bisphenol A (BPA) Concentrations."

87 For more of this history, see Meikle, *American Plastic.*

88 For a supporting argument about how actions like reducing plastic use or increasing recycling do not in fact cede access to land, see MacBride, "Does Recycling Actually Conserve or Preserve Things?" Spoiler: not really, no.

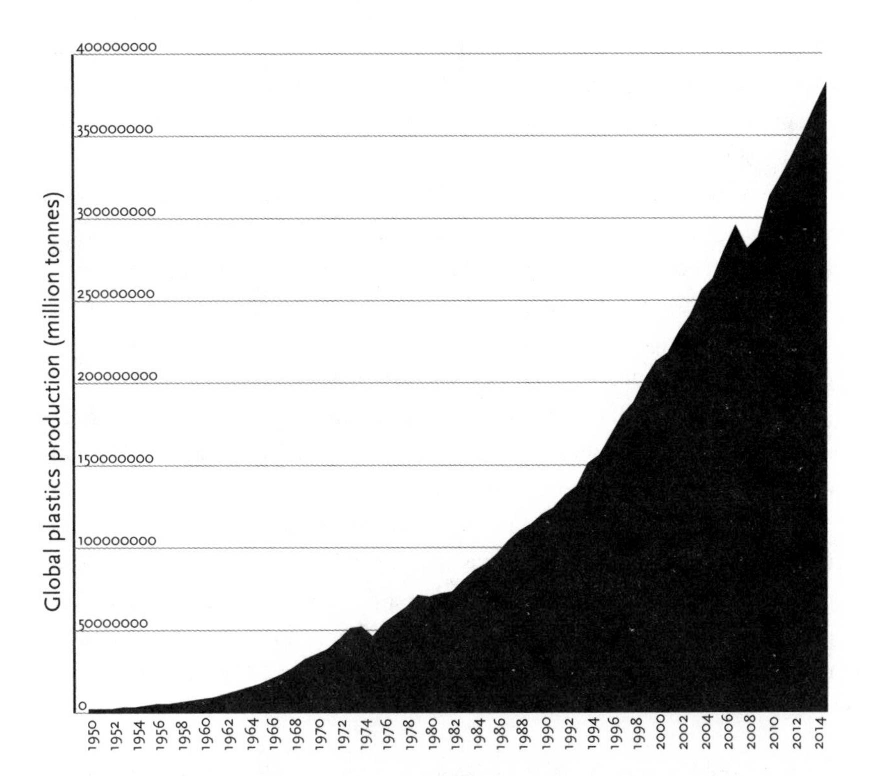

FIGURE 2.2. Increase of global plastic production, measured in tonnes per year, from 1950 to 2015. In 1950 the world produced only two million tonnes of plastics per year. Since then, annual production has increased nearly two-hundred-fold, reaching 381 million tonnes in 2015. For context, this is roughly equivalent to the mass of two-thirds of the world's population. Ritchie and Roser, "Plastic Pollution." CC-BY 4.0.

that they could address these plastics, they would also be eating the plastics in bridges, airplanes, automobiles, pacemakers, and buildings. Everyday infrastructure would be crumbling around us, like a B horror movie. Plastics undergird and prop up most urban, many rural, and even bodily infrastructures. They literally support life.

On a trip across the North Atlantic gyre, colleagues (all settler/unmarked) and I came across fishing ropes and a buoy snarled into a ball. It acted as a fish aggregator, providing shade for larger fish and a platform for micro-algae and microbes to grow, hosting a small but thriving ecosystem. The fish beneath the plastics were species that usually live close to land, among coral, but the plastics had provided a similar habitat thousands of miles from shore. Using scanning electron microscopy and gene sequencing, scientists have found more than one thousand types of bacterial cells on ocean plastic samples.[89] Tiny plastic pieces are home to plants, algae, and bacteria, the animals that feed on them, the predators that feed on these, and other organisms that establish synergistic relationships.

Scientists call this unique ecosystem "the plastisphere." Sitting in our boat, we faced an ethical question: do we take the tangled plastics out of the water, killing the life on and around it, or do we leave them in as the supporting structure of a functioning ecosystem?[90] Are the animals full of EDCs? Probably. Do the chemicals associated with plastics have health effects? Likely.[91] Were the individuals and overall ecosystem alive and thriving? Yes.

I have lost count of the number of academic presentations, usually in the humanities and social sciences, that use artist Chris Jordan's (unmarked) photographs of albatross carcasses on Midway Atoll with plastics in their rotting guts. The assumption in these presentations is always that the birds have died from ingesting the plastic, and the point is always that "the earth"[92] is in peril. But albatross do not tend to die from ingesting those plastics. Scientists noticed the plastics in albatross long before Chris Jordan did, and they have been studying this for decades. Here's one finding: "In a study on Sand Island, Midway Atoll,

89 There is a wealth of both natural and social science writing on this, including but not limited to Zettler, Mincer, and Amaral-Zettler, "Life in the 'Plastisphere'"; Amaral-Zettler et al., "Biogeography of the Plastisphere"; Reisser et al., "Millimeter-Sized Marine Plastics"; H. Davis, "Toxic Progeny"; and De Wolff, "Plastic Naturecultures."

90 My colleagues chose to pull it out. It was, after all, their mission to rid the seas of plastic. I think if CLEAR member Nicole Power (settler), who specializes in animal relations in science, had been there, we might have had a more robust discussion and maybe a different outcome. Thank you, Nicole, for your ongoing and generous intellectual companionship, in and out of CLEAR. Your everyday activism as a rural, settler, feminist killjoy is inspirational and so very necessary. Thank you for doing that tiring and crucial work, as well as for the investment you've made in my intellectual and political teachings about what it is to be a researcher on the island of Newfoundland. CLEAR and I are lucky to have you. So are Grandmother and Kookum. Thank you.

91 Scientists Lavers et al. (unmarked) argue, "Current knowledge of the negative effects of debris on wildlife is largely based on consequences that are readily observed, such as entanglement or starvation. Many interactions with debris, however, result in less visible and poorly documented sublethal effects, and as a consequence, the true impact of plastic is underestimated." Lavers, Hutton, and Bond, "Clinical Pathology of Plastic Ingestion in Marine Birds."

92 There is an extensive literature on the production of "the global" as a particular scale, of which these arguments are part, including Edwards, *Closed World*; Masco, "Bad Weather"; and Dunaway, *Natural Visions*.

in 1987, no Laysan Albatross chick deaths, impactions or ulcerations in proventricular [stomach] linings were attributed to ingested plastic."[93] Here's another. In a fourteen-year study of plastic ingestion by seabirds, scientists "found no evidence that seabird health was affected by the presence of plastic, even in species containing the largest quantities" and "the [study] results are not evidence of a cause-and-effect link between plastic ingestion with chick death. While it is possible that the death of healthy chicks may result from ingested plastic, it is also possible that unhealthy chicks eat greater amounts of plastic from the ground as a result of their poor condition."[94] While "seabird [populations] are declining faster than any other group of birds"[95] in the world, "the Laysan albatross population increased from an estimated 18,000 pairs in 1923 to 590,000 pairs in 2005"[96] on the Midway Atoll. They are now considered one of the most successful seabird species in the world. It's believed that they have reached their carrying capacity for the islands where most of the photos of their carcasses come from.[97]

So why all the dead albatross? It's natural, even Natural. Albatross are categorized as a k-selected species, as are humans, meaning they have relatively stable populations characterized by members that live to be quite old and produce a few young every year, most of whom die. This high rate of death for young albatross is normal. For young albatross on Midway Atoll, "morbidity can be substantial with 1000+ chicks dying per day."[98] This has happened since time immemorial, since before plastics. It's just that now, when they die, they have also ingested plastics. There is no scientific evidence for causal effect between the two. This does not mean that their ingestion of plastics doesn't involve types of harm that exceed science or that plastics are totally fine for birds to eat, but that ingested plastics are not the determining factor in these birds' deaths. To conflate bad relations (plastics in bellies) with scientific harm (the cause of mortality) not only misidentifies which relations matter, but also, from a scientific per-

93 Sileo, Sievert, and Samuel, "Causes of Mortality of Albatross Chicks at Midway Atoll."

94 Auman et al., "Plastic Ingestion by Laysan Albatross Chicks on Sand Island, Midway Atoll." These are just two of a dozen studies on albatross on the Midway Atoll, and all tell an identical story.

95 Lavers and Bond, "Ingested Plastic as a Route for Trace Metals in Laysan Albatross (*Phoebastria immutabilis*) and Bonin Petrel (*Pterodroma hypoleuca*) from Midway Atoll," 493.

96 Arata, Sievert, and Naughton, *Status Assessment of Laysan and Black-Footed Albatrosses*, 2. Yeah, that's *a lot* of birds. Midway Atoll is the species' main breeding and nesting spot.

97 Arata, Sievert, and Naughton, *Status Assessment of Laysan and Black-Footed Albatrosses*, 2.

98 Work, Smith, and Duncan, "Necrotizing Enteritis as a Cause of Mortality in Laysan Albatross."

spective, misidentifies where albatross *are* facing mortality by plastics—which is via bycatch from line fishing.[99]

That means that the iconic photos of albatross, which ingest some of the highest rates of plastics of any species, are images of survival and success, not peril and doom. Albatross are all about Murphy's alterlife, "the condition of being already co-constituted by material entanglements with water, chemicals, soil, atmospheres, microbes, and built environments, and also the condition of being open to ongoing becoming." They are the life that holds "together tensions between violence and possibility, braiding the organic and inorganic, body and land" and thus represent an "openness to a potential for recomposition that exceeds the ongoing aftermaths."[100] In short, albatross are effing stars. To use albatross bodies as tokens of damage instead of signs of alterlife is not only incorrect and a missed opportunity, it is rude.[101] It misses the wider relations, the Land relations, of albatross and plastics, and turns them into a Resource for shock, awe, and charismatic academic presentations. Please, stop. Thank you.

Plastics Are Land

Elizabeth Hoover has conducted research with people in the Mohawk community of Akwesasne about their fish consumption practices in the face of ongoing advisories recommending that people not eat fish contaminated with polychlorinated biphenyls (PCBs). PCBs are EDCs that have historically been used as flame retardants in plastics and other materials. Hoover explains that some people choose to eat the fish regardless of the advisories: "We give thanks for that food and we have to use it. . . . I mean it doesn't make sense scientifically, but it makes sense spiritually and mentally that you should eat that, you know. You

99 Arata, Sievert, and Naughton, *Status Assessment of Laysan and Black-Footed Albatrosses*.

100 Murphy, "Against Population, towards Alterlife," 118.

101 When I say rude, I mean that it erases the considerable agency of albatross. But also, the majority of the social science and humanities presentations that I have seen that include these pictures also tend to make claims about extending kinship to "nonhumans." They almost always mean albatross and almost never mean plastics. There are multiple levels of rude happening here: choosing your kin instead of being invited in and chosen by kin (which is a sign of Whiteness according to Aileen Moreton-Robinson and Darryl Leroux), excluding undesirable kin while simultaneously extending the definition of kin, using Indigenous (most often Cree and Métis) concepts of kin without citation (see Todd), and, finally, putting a picture of your dead and rotting "kin" on the board. Seriously. Rude. JFC. Maybe you don't mean *kin* after all. Moreton-Robinson, *White Possessive*; Leroux, *Distorted Descent*; Todd, "Indigenous Feminist's Take on the Ontological Turn."

can't just put it aside and say, 'well your [the fish's] work is not good enough,' or something, you know? They're still given out what their original instructions were, and it's us that are at fault, it's our fault that they're like that, you know."[102]

In the context of these relations and obligations to Land, it makes sense to some people to eat contaminated food. The stakes of not eating the food are cultural genocide,[103] where the languages, practices, knowledge, and thus relations with Land are killed[104] to the point that they are no longer reproduced by successive generations. In an interview between Hoover and Henry Lickers (Mohawk) at Akwesasne, Lickers talks about how the language, culture, knowledge, social relations, and practices surrounding tying knots in fishing nets are interrelated, and how these things can be erased when fish consumption advisories make the nets useless: "People forget, in their own culture, what you call the

102 Hoover, "Cultural and Health Implications of Fish Advisories," 6.

103 To be clear, cultural genocide is genocide. It's not genocide-lite. The use of the term *cultural genocide* here designates the particular mechanism of genocide. The term *cultural genocide* was popularized in Canada following the publication of the Truth and Reconciliation Commission (TRC) report in 2015, which documented the stories and testimonies of the survivors of residential schools in Canada. Residential schools were government-funded, church-run schools specifically designed to eliminate familial and community involvement in the intellectual, cultural, and spiritual development of Indigenous children. It was a way to "kill the Indian, save the man." The TRC made extensive use of the term *cultural genocide* in describing the residential schools, a controversial decision not least because the Canadian settler state was using it to avoid the definition of genocide under the United Nations Genocide Convention of 1948. This footnote is wearing a T-shirt that says, "Cultural genocide is genocide" on the front and "The Canadian settler state is a fuck nugget" on the back. Truth and Reconciliation Commission of Canada, *Honouring the Truth, Reconciling for the Future*; MacDonald, "Five Reasons the TRC Chose 'Cultural Genocide'"; Schwartz, "Cultural Genocide Label for Residential Schools Has No Legal Implications"; Staniforth, "'Cultural Genocide'?"

104 Indigenous languages aren't lost. We don't lose languages like we lose car keys. Like maybe we'll find the Beothuk language between the couch cushions. Languages are killed. Dr. Nicole Powers (settler) has taught me that cultures, whether they are rural Newfoundland cultures or Indigenous ones, do not *die out* because of a slow dwindling away of young people and jobs, or of knowledge holders and Elders. Rather, they are actively and systemically *killed*. They are killed by structural violence, often state violence, that ensures that continuing to live in a place becomes difficult, undesirable, or impossible. Or that speaking a particular language will bring violence or cause you to lose access to things you need, which is exactly what the residential schools were for. This is particularly appropriate in the case of pollution, when the settler state creates and supports threshold levels of allowable pollution.

knot that you tie in a net. And so, a whole section of your language and culture is lost because no one is tying those nets anymore [if people don't eat the fish]. The interrelation between men and women, when they tied nets, the relationship between adults or elders and young people, as they tied nets together, the stories . . . that whole social infrastructure that was around the fabrication of that net disappeared."[105]

These interviews show how Land relations do not separate out like relations with Nature. As Hoover argues in *The River Is in Us*, "The greatest health effects [of pollution] are seen outside of chemical exposures and are thus not included in risk assessments. The environmental contamination in Akwesasne has negative impacts on the cultural, social, and physical health of the community beyond those directly related to the ingestion of fish."[106] Here, pollution, language, eating, and obligation are part of the same bundle. This not only changes what a chemical is, but also transforms what appropriate responses to contamination look like.

Land relations—including colonial ones!—are alterlife relations. The question is what kinds of lives and lifeworlds are being reproduced in those relations, and which are not. Murphy calls this "reproductive justice." She reminds us, "Reproduction itself is not a good; rather, it is a process of supporting some things and not others,"[107] and as such it is important to "rework reproduction to conceptualize how collectivities persist and redistribute into the future and to query what gets reproduced," including "the uneven relations and infrastructures that shape what forms of life are supported to persist, thrive, and alter, and what forms of life are destroyed, injured, and constrained."[108] As such, "Reproductive justice is the struggle for the collective conditions for sustaining life and persisting over time amid life-negating structural forces, and not just the right to have or not have children. Reproductive justice is thus inseparable from environmental justice, antiracism, and anticolonialism."[109]

This is why using photos of albatross to denote destruction rather than the presence of sustained and persisting life is so rude. This is why eating contaminated fish is life (as is not eating contaminated fish, from another point of view).

105 Hoover, "Cultural and Health Implications of Fish Advisories," 4.
106 Hoover, *River Is in Us*, 215.
107 Murphy, *Economization of Life*, 142.
108 Murphy, *Economization of Life*, 141–42.
109 Murphy, *Economization of Life*, 142.

Purity is not an option here—plastics are already in Land relations. Shawn Wilson (Cree) writes about an Elder discussing his computer: "He went on to say: This machine here is made from mother earth. It has a spirit of its own. This spirit probably hasn't been recognized and given the right respect that it should. When we work in a world of automated things, we forget that . . . everything is sacred, and that includes what we make."[110] Zoe Todd (Métis) similarly writes about oil as Land: "We may go the way of the dinosaurs, and it will be because the dominant human ideological paradigm of our day forgot to tend with care to the oil, the gas, and all of the beings of this place. Forgot to tend to relationships, to ceremony (in all the plurality of ways this may be enacted), to the continuous co-constitution of life-worlds between humans and others."[111] Both Wilson's Elder and Zoe Todd are also talking about plastics, since computers are made of plastics, and since oil and gas are the raw feedstock for plastics and many of their associated chemicals.

My use of Wilson's, Todd's, and Hoover's words to show how plastics and toxicants are Land is not an invitation for settlers to put "Plastics Are Land" on their PowerPoint slides instead of those pictures of albatross. Indigenous Land relations, delicious as they may be for "thinking with" or "drawing upon," are not for consumption or appropriation by settlers.[112] In earlier drafts of this chapter, I had framed the discussion around the thesis that "Plastics Are Kin," but I changed this after conversations with various Indigenous thinkers and Elders about complex issues of kinmaking with bad kin and the already rampant fe-

110 Wilson, *Research Is Ceremony*, 90.

111 Todd, "Fish, Kin and Hope," 105.

112 If you aren't sure about whether or not your research is appropriating, the Aorta Collective has an amazing resource that includes questions to ask yourself, such as "Does the source group or culture have a history of exploitation, slavery, or genocide? If so, there is already a social power dynamic at play regarding the use of their culture. Are the people/the culture from whom this imagery, item, or custom comes benefitting? Has the source community invited you to share in this?" My personal favourite is "Is the source's significance filling a hunger (for 'sacredness,' for 'meaning')?" Kin are not chips. See Anti-oppression Resource and Training Alliance, "Cultural Appropriation: Guiding Questions to Ask," April 2015, http://aorta.coop/wp-content/uploads/2017/12/Cultural-Appropriation-Resource-Sheet-April-2015.pdf.

Thank you, AORTA Collective (Anti-oppression Resource and Training Alliance), for the considerable gifts you share. Your work on anti-oppressive facilitation is the cornerstone of my facilitation practices, including in CLEAR, and the way you articulate your theory of change as a justice organization is exemplary. Thank you for your generosity, intelligence, and guidance.

tishisation of nonhumans as kin by academics as acts of possession and redemption. I removed conversations that were not fit or ready for public consumption, changing the chapter to "Plastics Are Land." But even that left a lot of room for creepiness, so I instead reframed the chapter around scale as relationships that matter, with only a small introduction to Land-plastic relations with an Indigenous frame, here.

When I say creepy, I'm not being glib. Creepiness is a relation directed by the intense desire of one party toward another, with that desire so obfuscated, unknowable, or such a bad fit that the originating desires do not quite make sense to the object of desire and can even constitute violation. The increasingly popular academic conversations among and emanating from settlers about kin and Land are hella creepy. I can never tell what most people mean by kin or Land, especially because both are usually positioned as inherently good[113] (which is weird if you have any experience with family members or weather, to name two obvious manifestations of kin and Land that can be monumentally shitty and even dangerous). This is why I offer scale: it scaffolds a lesson on the specificity of relations that larger concepts of kin and Land require without easily slipping into fetishism or claiming someone else's Land relations. Hopefully the concept of scale as an analytic to suss out which relationships matter in different contexts is useful both to readers who have strong desires to understand kin and Land but not a lot of teaching on the topics, as well as to readers who have lived in these lessons through community for generations and are well-versed participants in Land and kin.

We do not need to appropriate Indigenous Land relations as a Resource to think about and then enact anticolonial land relations—of which there are many. When I say plastics are Land, I lean on Wilson, Todd, and Hoover to show how this is already possible and in fact common in some worlds, but this is not the only way that plastics can be in anticolonial land relations. Different L/lands live together all the time—not always well, but they do.[114] Knowledge systems such as political ecology, cultural geography, and environmental justice are just some of the ways to look at how systems of value and knowledge animate relations. Scale is another. To return to the discussion from this book's

113 For an exception to this trend, which looks at the harms and violence of kinship while still foregrounding obligation and ties, see Fennell, "Family Toxic." I found this text while trying to add nuance to academic discussions that assume kinship is inherently good or One Big Happy F.

114 This insight and phrase come from a discussion with Emily Simmonds (Métis). Thank you, Emily.

introduction on the problem of a universal, humanwide "we," the lessons from plastics and their myriad relations at different scales, along with the many different lands that exist in tandem with them, afford us a "focus on responsibility—the obligation to enact good relations as scientists, scholars, readers, and to account for our relations when they are not good. And you can't have obligation without specificity."[115] Lands (with or without capitalization) and relations (whether understood from Indigenous worldviews or not) are always about specificity, about the relations that matter in each context. Where are plastics in the various land relations around you, Reader? What can plastics teach you? What is a chemical where you are?

115 As I stated in the introduction.

3 · An Anticolonial Pollution Science

Every morning when I put on my lab coat, I have decisions to make. How will we do science today? How will we work against scientific premises that separate humans from Nature, that envision natural relations as universal, and that assume access to Indigenous Land, especially when so much of our scientific training has primed us to reproduce these things? These are not theoretical questions—they are practical questions, questions of method-and-ethics (hyphenated because they are the same thing). Critique is important[1] but it can only take you so far when you are a practitioner trying to do work in a good way.

In *Molecular Feminisms*, Deboleena Roy (diaspora Indian)[2] recounts how her commitments to the laboratory presented "challenges of actually trying to apply feminist epistemologies and methodologies at the level of practices at the lab bench."[3] Roy chose to "address some big questions that both feminist scientists and scientist feminists may have in common. How do we continue with science after the critiques of science? How do we work toward biology that we desire? How are we to encounter matter? How can we bring questions of context with us when we do encounter this matter? How can we reconfig-

1 For an argument about the necessary place of critique beside action, see Hale, "Activist Research v. Cultural Critique."

2 See the introduction, footnote 10, for more on these designations. In short: rather than leaving (usually) white and settler authors unmarked as the unexceptional norm that does not have to introduce itself, I mark everyone with either how they introduce themselves in their texts or as unmarked if they do not introduce themselves based on their L/land relations or relationships to privilege. You might start to notice a pattern . . .

3 Roy, *Molecular Feminisms*, 11.

ure the relationship between the scientific knower and what is to become the known?"[4] Likewise, feminist scientist Banu Subramaniam (Indian)[5] asks how to bring about the goal of feminist science studies to "develop an experimental practice and method that does not overdetermine or prefigure its conclusions" and instead makes room for "imagination and gusto," reflexivity and reconstruction, in experimental biology.[6]

Roy and others show that a commitment to doing science in a feminist way places critique in a unique relationship to scientific method. As feminist geographers Nadine Schuurman (unmarked) and Geraldine Pratt (unmarked) write,

> "How" critique is expressed, as well as what its objectives are, is critical to achieving changes in any research area. We start from the position that many of the critiques of Geographic Information Systems (GIS) have aimed to demonstrate what is "wrong" with this subdiscipline of geography rather than engaging critically with the technology. Critics have judged the processes and outcomes of GIS as problematic without grounding their criticism in the practices of the technology. This follows a pattern of external critique in which the investigator has little at stake in the outcome. External critiques . . . tend to be concerned with epistemological assumptions and social repercussions, while internal critiques have focused on the technical. But there is a further difference. Internal critiques have a stake in the future of the technology while external ones tend not to. . . . We argue for a form of critique that transcends this binary by tackling enframing assumptions while remaining invested in the subject. To be constructive, critique must care for the subject.[7]

Care for the subject of critique is part of feminist methodologies.[8]

The most useful definition of care I've heard was articulated by CLEAR

4 Roy, *Molecular Feminisms*, 12.

5 The method of introducing people gets meta with this one: the section *Ghost Stories for Darwin* where Subramaniam introduces herself in various ways is about how she was taught to "encode cultural difference within the language of meritocracy" and understand racialization, gendering, and heteronormativity as part of scientific subjecthood and the tensions of having to/not/un-identify with aspects of that subjecthood. I did say these identifications are an imperfect method. How you introduce yourself changes with your audiences and this practice makes that static. Subramaniam, *Ghost Stories for Darwin*, 174 and 171–79.

6 Subramaniam, *Ghost Stories for Darwin*, 4, 5.

7 Schuurman and Pratt, "Care of the Subject," 291.

8 TallBear, "Standing with and Speaking as Faith." More on this is in the introduction.

member Emily Simmonds (Métis) during a lab meeting.[9] We were talking about killing animals for science and how to kill in a good way. Simmonds spoke of care as an affective relation whose leading ethic is to create attachments within infrastructures of inequity. These attachments are best described as obligations. What I like about this working definition is that it allows things like genocidal residential schools to be about care. Missionary care was often well intentioned, part of the "save the man, kill the Indian" Christian and settler state logics of colonial paternalism and annihilation.[10] They certainly made (violently) affective relations that made (blistering) attachments in infrastructures of (colonial, genocidal) inequity that the schools understood as their (Christian) obligation. From the position of conquest (of people, Land, and souls), genocidal care is an obligation. As feminist scholars Aryn Martin (unmarked), Natasha Myers (settler), and Ana Viseu (unmarked) point out, "Practices of care are always shot through with asymmetrical power relations. . . . Care organizes, classifies, and disciplines bodies. Colonial regimes show us precisely how care can become a means of governance."[11] Care is not inherently good.[12] It is an uneven relation and can contribute to and/or mitigate unevenness.

This is a crucial framing for attempting to change dominant science[13] while wearing a lab coat. All science has L/land relations, as discussed in the previous chapters. Some of these relations are colonial and we have to maneuver within them: there is no blank slate to start from.

9 One of the many things Simmonds researches is how uranium economies produce and amplify colonial geographies, primarily through concepts of consent, colonial infrastructures, toxic sovereignties, and the biopolitics of settler colonialism. She thinks about care a lot. During the lab meeting where she articulated her working definition of care, we were talking about our animal respect protocols. Since we deal with dead animals, and often kill animals, many of the definitions of care from feminist STS were not working for us. This meeting occurred in 2018. This definition builds on much existing work in feminist STS, which foregrounds power, obligation, responsibility, and ethics in care, in particular: Murphy, "Unsettling Care"; A. Martin, Myers, and Viseu, "Politics of Care in Technoscience"; Puig de la Bellacasa, "Matters of Care in Technoscience"; Schrader, "Responding to *Pfiesteria piscicida* (the Fish Killer)."

10 Grande, *Red Pedagogy*.

11 A. Martin, Myers, and Viseu, "Politics of Care in Technoscience," 628. While all authors are unmarked in this text, Myers introduces herself and her critical standpoint as "a settler living and working on stolen Indigenous lands" in Evans, "Becoming Sensor in the Planthropocene."

12 Murphy, "Unsettling Care"; Ureta, "Caring for Waste."

13 See footnote 77 in the introduction on why I use the term *dominant science* instead of *Western science*.

I remember doing my first plastic ingestion study on dovekies, not as a CLEAR member, but training under ornithologists. Over two hundred dovekies had been wrecked in a storm, meaning they died as a group, likely after being blown into a cliff and then drowning in the water below. At first, I thought the tiny, red, narrow fragments I found in their bellies might be plastics (what else is that red in nature?!), but then I saw the larger branched structure they'd come from—some kind of seaweed. I learned to hold judgment on unfamiliar things until I went through more dovekie guts and thus more of the environment they lived and ate in. Those clear "plastic" films? Bits of shrimp exoskeletons. I could learn about underwater landscapes I'd never been to through the dovekies' gizzards. Plastics were part of those landscapes. Especially green fishing line.

The warm feelings I had from learning about underwater landscapes stopped short when suddenly it occurred to me to ask: how did we get these birds? Did [our collaborator] just take them? Did we get permission? Accessing landscapes, underwater or otherwise, suddenly seemed shitty.[14]

Civic Laboratory for Environmental Action Research (CLEAR) is the land-based, feminist, and anticolonial environmental science lab I direct with between five and twenty-five collaborators. In CLEAR, we mostly count plastics. These plastics are often in the gastrointestinal tracts of animals caught for food, but we also sample plastics in water, in sediments, in ice and snow, and on shorelines. Many pollution scientists create counts that make thresholds seem like properties of Nature; 15 units of pollution are fine but 16 is too many (see chapter 1). But there are also ways to count that further an "epistemology where the relationship with something is more important than the thing itself. Inherent in this concept is the recognition that this person, object or idea may have different relationships with someone or something else."[15] The rest of

14 The lab stories that run throughout this chapter as well as other parts of the book are from various members of CLEAR, including myself. All stories are shared with permission.

The permission to obtain dovekies is tricky when it comes to the area around St. John's where the birds were recovered. Normally you would get permission from the Indigenous groups whose Land the birds died on, but the Beothuk were completely murdered by white settlers during the conquest of the island of Newfoundland. No one speaks on behalf of the Beothuk, so permission can never be granted. This inability to follow basic protocol in the face of genocide is a problem, and frankly one of the goals and achievements of genocide.

15 Wilson, *Research Is Ceremony*, 73.

FIGURE 3.1. Contents of a dovekie's gizzard and proventriculus (area between the crop and gizzard) under a microscope. Note the branched structure on the lower right. To horrified scientists seeing the state of the stage: yes, this was a scavenged microscope, and no, we were not using it to identify microfibers (though there are at least two in this image). This was early days. Photo by Max Liboiron.

this chapter shows how this can happen in a marine laboratory that is well ensconced in dominant and Western science while also conducting anticolonial science.

I've been working as an apprentice with a group of ornithologists, picking plastics out of the bellies of starved dovekies. We process more than one hundred birds, but one dovekie surprises us: D-156 (our 156th dovekie). I call over the biology student working with me to witness the number of plastics I am pulling out . . . 7, 8, 9 . . . ! Until now, about a third of the birds have ingested plastics, but only 1 to 3 plastics on average. . . . 32, 33, 34! . . . We pull out 50 plastics. Oddly, many are burned. We talk about D-156 for days. Maybe she got into a campfire and ate up its ashes. Maybe she got separated and then united with her flock. Maybe she was the only one who chose to eat items that weren't food when she was starving. When it comes time to analyze the data, the rest of the research team call her an outlier and talk about leaving D-156 out of the study, both because of her high number of ingested plastics, but

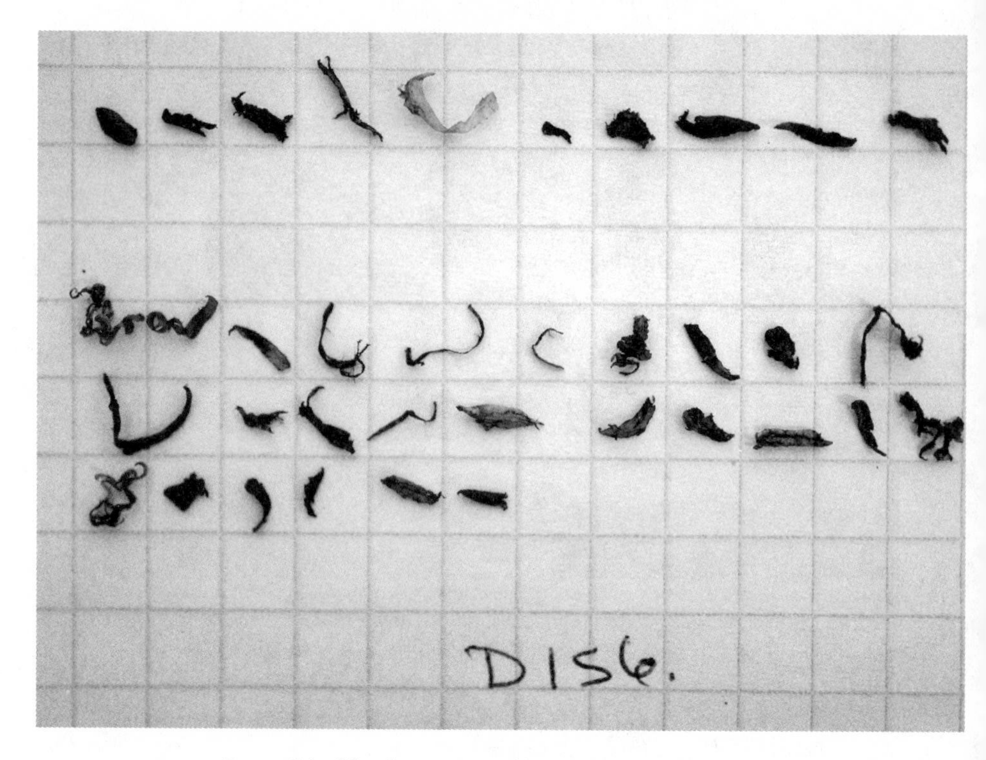

FIGURE 3.2. Some of the fifty plastics ingested by D-156. It is odd how they are all similar sizes and shapes. They likely came from the same place, and while we have many theories about what that place might have been (cruise ship garburator/incinerator, fishing-camp fire), there is no way to know. Photo by Max Liboiron.

also because of the odd burned plastics she ingested. I argue that D-156 is an extreme but quintessential example of plastic ingestion and while she had a different eating pattern than her flock, she was still very much part of that flock and its collective behaviours. D-156 stays in the final paper but is called an outlier.[16] *The burned plastics are mentioned in the paper but are separated from D-156 and her high plastic count so they stand alone, in aggregate and without context. D-156 isn't even named. Now that I run my own lab, we wouldn't have called her an outlier and likely would have given her more of a holistic portrait. Maybe like the one you just read.*

16 Avery-Gomm et al., "Study of Wrecked Dovekies (*Alle alle*) in the Western North Atlantic."

outlier-data

Making and Doing Anticolonial Science

The existence of D-156 presented us with a series of questions: Do we include the outlier, or not? Why? What kind of world do we describe either way, and to what ends? If we choose to include it, how? Do we use the statistical definition of an outlier, which depends on a normal curve, or do we note that D-156 was different? I'd never thought of outliers in this way before, though I was a feminist the whole time. Through our handling of dead birds, little red organic or plastic bits, specimen bags, tweezers, and the statistics of outliers, we find that we must be accountable to these things and their worlds in ways that don't always show themselves when we are theorizing at our desks and handling keyboards and books. Thinking at desks is still a way of doing, of course.[17] But when your hands are in someone's guts unanticipated issues tend to present themselves.

How?

Understanding accountability in practice *through* practice is a core strength of "making and doing" as a methodology in science and technology studies (STS), "a mode of scholarship that involves attending not only to what the scholar makes and does but also to how the scholar and the scholarship get made and done in the process."[18] The question of how is CLEAR's main concern. Scientists count plastics in animal guts with some regularity—*how* will we do it? Statistics happen every day—*how* will we do ours? And *how* will they be in good L/land relations? *How* will we do science in an anticolonial way, rather than merely with anticolonial intent?

This question of *how* is common in STS, which often asks how, exactly, is knowledge made? How does laboratory knowledge come to be? Through what inscriptions, conversations, or bumps against the machine? I don't mean this kind of *how*.[19]

For CLEAR, we mean to ask *how* the way Vanessa Watts (Anishinaabe and

17 See chapter 2, right around the text with footnote 110, where an Elder is talking about his computer as a relation, as a thing with spirit that we collaborate with rather than control and use.

18 Downey and Zuiderent-Jerak, "Making and Doing," 225.

19 The genre of study called "laboratory life" or "laboratory studies" was popularized by the French STS scholar Bruno Latour (unmarked), who is also really into nonhumans like door hinges. I'm pretty sure CLEAR's focus on *how* is not what Latour means when he follows scientists around for their how. Latour and Woolgar, *Laboratory Life*.

While I draw on these kinds of studies, I also depart from them in significant ways by foregrounding obligation, and specifically obligation to L/land as the primary frame for

Haudenosaunee) means it. She talks about the way dominant settler understandings of agency and worldmaking practices remove the *how* and *why* out of the *what*. The what is left empty, readied for inscription. . . . The man-made [*sic*] distinction between what and how/why is not an innocent one"[20] because it leaves the discovery of the how in the human domain. This is true even in what are called "multispecies" and nonhuman encounters in academia, "meaning that, although the dirt/soil has been granted entrance into the human web of action, it is still relegated to a mere unwitting player in the game of human understandings."[21] Like a pet, always loved but certainly in deficit. These are Nature relations that maintain separation, rather than Anishinaabe and Haudenosaunee Land relations (see chapter 1). Damn, I love that article.

How as Accountability

We mean *how* the way Shawn Wilson (Cree) means it when he says "the shared aspect of an Indigenous ontology and epistemology is relationality (relationships do not merely shape reality, they are reality). The shared aspect of an Indigenous axiology [ethics] and methodology [doing] is accountability to relationships."[22] Here, *how* is not a process. Careful. This is the tricky bit. *How* is a genre of relationality based in obligation.[23] As educator Dwayne Donald (Cree) explains, "This form of relationality is . . . an ethical stance that requires attentiveness to the responsibilities that come with a declaration of being in relation."[24] In a cosmology based on relationality-as-accountability, Wilson reminds us that "right or wrong; validity; statistically significant; worthy or un-

---

scientific how. There are laboratory ethnographies by outsiders more aligned with this approach, such as Helmreich, *Alien Ocean*.

20 Watts, "Indigenous Place-Thought," 24; emphasis in original.

21 Watts, "Indigenous Place-Thought," 30.

22 Wilson, *Research Is Ceremony*, 7.

23 Some Indigenous reviewers (thank you!) have asked me, why do you always say obligation instead of responsibility? As I understand it, responsibility is a choice (you can take responsibility, or not), whereas obligation precedes you and is not a choice (you are obliged even if you don't take responsibility). For a mixed audience, I choose the heavier hand. If responsibility is in your or your Elders' vernacular, go for it. Terms like *accountability*, *obligation*, and *responsibility* have been heavily co-opted and will have different baggage for different folks. The way I understand obligation is that it is another way of saying gratitude, and gratitude is your gift to the world (water, relatives, food, sunlight, or Land for short) that gives you more than you can ever return. But I could be wrong.

24 Donald, "Indigenous Métissage," 535. Also see Donald, Glanfield, and Sterenberg, "Living Ethically."

worthy; value judgements lose their meaning. What is more important and meaningful is fulfilling a role and obligations in the research relationship—that is, being accountable to your relations."[25] This does not mean that relations exist and you are accountable to them through your actions, but rather that things are constituted by these relations (as articulated in much science and technology studies) and that accountability is the way to describe that constitution (which is common in many Indigenous theories).

When I think about the *how* of our science, I think about something my mentor and Elder Rick Chavolla (Kumeyaay) recalls his mother saying, "taking a chocolate bar out of your back pocket is a prayer." It's not that she loved chocolate in a holy way. (I think) she is saying that everything you do is a prayer, where prayer shows and reinforces our obligations and gratitude to Land. CLEAR member Edward Allen (Kablunangajuk) explains it another way: "Ceremony is about teaching and learning, and it reinforces and perpetuates what is meaningful to us. Ceremony can be prescriptive or a regular part of the maintenance of our well-being. With the danger of oversimplifying it, ceremony is an enactment of our values, guiding principles, and our prayers. Our prayers are the acknowledgment of what is sacred, and what is sacred is how we are connected to everything else"[26] . . . including back pockets and chocolate bars, Excel sheets and lab benches. This is what we mean by the *how* of science. We mean making and doing ceremony in science.

The other part of *how* is its undoing. As la paperson writes, "'How?' is a question you ask if you are concerned with the mechanisms, not just the motives, of colonization. Instead of settler colonialism as an ideology, or as a history, you might consider settler colonialism as a set of technologies—a frame that could help you to forecast colonial next operations and to plot decolonial directions."[27] As chapter 1 illustrated, scientists don't have to be racist or intentionally imperial to reproduce and enforce colonial land relations that use Land as a Resource, flattening and hoarding its relations for colonial goals while maintaining the violent erasure of Indigenous relations and bodies; Streeter and Phelps simply made some measurements that showed how rivers can assimilate pollution. So, too, with anticolonialism. It can also be understood as a set of technologies, or even protocols, that make different Land relations. Because colonialism is ongoing and must be maintained, these mechanisms are a crucial way to think about anticolonial work. They are at the core of CLEAR's theory of change.

25 Wilson, *Research Is Ceremony*, 77.

26 Edward Allen, personal communication, August 17, 2016.

27 paperson, *A Third University Is Possible*, 5.

Protocol

To centre *how*, CLEAR focuses on L/land relations at the scale of protocol. Yes, we think about (and survive) genocide, missing and murdered Indigenous women and girls, and land theft, but when you're standing at the bench with a beaker in your hand, those questions are hard to bring to earth. Protocol helps with this. When I say protocol, I use its double meaning in both science and ceremony to mean "the manner in which one approaches each and every element in our space"[28] as a manifestation of our values, survival, and goals, as axiology-in-practice. In a scientific laboratory, protocols are the scripts you follow to keep your controls controlled, your science replicable, and your findings valid. Step 1: Tie back your hair and put on gloves to avoid contamination. Step 2: Rinse the outside of the specimen bag in water before placing it in the sieve. Protocol also refers to guidelines for conduct during ceremony: bring the hosting Elder tobacco (loose cigarette tobacco will do, but leaf tobacco is better) in a red cloth bundle for the paarantii kaayash ooshchi;[29] present it in your open left hand and let the Elder take it from you. In both science and ceremony, protocols reinforce and perpetuate what is meaningful and right in an activity.[30]

Protocol can manifest in small ways. In one of CLEAR's protocols, for instance, we do not wear earbuds or headphones when we dissect fish, since they are L/land and it's rude to tune out your relations. Sometimes protocol manifests in more notable ways, like redirecting hundreds of thousands of dollars of federal grant monies to Indigenous-led research instead of settler-led research on Indigenous Land.

Feminist scholar Helen Longino (unmarked) proposes that "we focus on science as practice rather than content, as process rather than product, hence, not on feminist science, but on doing sciences as a feminist."[31] So, too, with anticolonial science, where we focus on doing science with an orientation to good L/land relations. The thing about protocols is that they are orienting technologies, pointing us toward certain futures that are good and right and true, rather than merely describing a series of actions or processes. The following protocol is excerpted from CLEAR's lab book:

28 Keali'ikanaka'oleohaililani, "Hawaii Environmental Kinship," 77.

29 Question: Hey, why don't you italicize the Michif? Answer: Because you italicize foreign languages and that would be English, not Indigenous languages. Italicizing a whole book minus these lines would be annoying. But I did think about it.

30 Whyte, Brewer, and Johnson, "Weaving Indigenous Science"; TallBear, "Standing with and Speaking as Faith."

31 Longino, "Can There Be a Feminist Science?" 52.

Processing the stomach:

1 Do not wear earbuds to listen to music while processing, as this separates you from the animal, who deserves your full attention and respect. You can play music from a speaker, and singing is particularly welcome.
2 Take a moment to think about the samples and where they came from.
3 Fill in the spreadsheet with the fish code (ex. PH13, NCCED18–01), today's date, the location the fish was caught, size, and sex if it is not already filled in. This will require you to look at the sample collection sheets or other documentation. Fill in your name, the date on the contamination control, and how you are feeling.[32]
4 Before opening each gut, wash your hands, backwash the sieve, and wipe down the tools, microscope lens & plate, and Petri dishes you will use. This will mitigate (not eliminate!) microfibers that have settled on tools through atmospheric deposition.
5 Stack the wide-grid sieve (if processing big guts) on top of the 5mm sieve on top of the 0.425mm sieve in the sink. The top sieve will catch the larger items and make visual inspection of the finer sieves easier. [Teaching moment: notice it says "finer" instead of "smaller" sieve? In science, it is important to get descriptions of relative quantities correct, so that size is not conflated with mass, mesh size, duration, etc. since they mean different things, have different relations.]

Even before we touch the fish guts, there are already several moments of orientation in these few moments of protocol: think about the fish, the land, and your relation to them. You don't have to be kin with the fish (though some of us are), but neither should you be thinking of the fish primarily as a specimen or scientific object. While the protocol asks a lab member to consider the fish or rinse the sieve, the lab member is also expected to think of other ways to respect the fish and reduce contamination—to become attuned to these relations and com-

32 Adding feelings to our data entry is a relatively new protocol for CLEAR and came about when we read *Data Feminism*, which highlights how data is often disembodied. Not only does our data entry and work attempt to highlight fish bodies but also the bodies of those doing the data entry. While we're not quite sure what we're going to do with this data, I've found it to be a source of surprise and generosity when I'm going through lab data and I see students are happy, struggling, or bored. It helps me take care of lab members (including reminding them to go home when they're sick) and helps us figure out where our practice and protocols are bogging down. D'Ignazio and Klein, *Data Feminism*.

port themselves accordingly, extending the protocol into new spaces to uphold the spirit of the script.

Whatever the scientific or ceremonial paradigm, protocols are enactments of our values and guiding principles, and they instruct us in how to reproduce what is good, whether that good is objectivity (sigh[33]) or good L/land relations, whether you're a settler with land relations or an Indigenous person with Land relations or something else. Sometimes protocols are prescriptive, and sometimes they are about the maintenance of everyday life, but they are always orienting you toward a particular horizon and away from others. They are reproductive technologies (see chapter 1).

Indigenous Sciences Are Different Than Anticolonial Sciences

"But I'm a settler! I can't do back-pocket chocolate-bar prayers! That's appropriation!" Good eye. Love you. I have been using Indigenous studies and science and technology studies scholarship from Indigenous writers to talk about a different orientation to science. By doing so, I've somewhat mushed together anticolonial science and Indigenous science though they are two different things.

Indigenous Sciences

Indigenous sciences are done by Indigenous peoples, full stop: "Native science is a metaphor for a wide range of tribal processes of perceiving, thinking, acting, and 'coming to know' that have evolved through [our collective] experience with the natural world."[34] Sometimes Indigenous sciences use methods, tools, theories, and frameworks developed out of Western and other non-Indigenous sciences, like the work of Robin Wall Kimmerer (Potawatomi).[35] Sometimes not. Sometimes they involve settler scientists. Sometimes not. Sometimes it is called Traditional Knowledge. Sometimes not. These decisions are an expression of Indigenous sovereignty over Indigenous ways of producing knowledge on Indigenous Lands, by Indigenous peoples.[36]

CLEAR does not claim to do Indigenous science, not least because most of

33 For a primer on how objectivity is a value-based concept that changes over time as Western societal values change, see Daston, "Objectivity versus Truth."

34 Cajete, *Native Science*, 2.

35 Kimmerer, *Braiding Sweetgrass*.

36 For more, see Geniusz, *Our Knowledge Is Not Primitive*; Kimmerer, *Braiding Sweetgrass*; Kawagley, *Yupiaq Worldview*; Kawagley, Norris-Tull, and Norris-Tull, "Indigenous

our members are white settlers. While some of our Inuit, Métis, and First Nations members certainly draw on Traditional Knowledge or local knowledge and certainly work from their worldviews and even with their families, communities, and homelands, we do not give this to academia.[37] Stacey Ann Langwick (unmarked) writes about a similar refusal within an NGO in Tanzania, where health clinics do not move their medical practices to Indigenous science even if practitioners' identities and knowledges might allow them to do so. She writes about how dawa lishe, a medical practice, "is not a return to, or even a nostalgia for, traditional African healing. It is, however, a refusal to forget in the present that African healing has long addressed humans and their environments together. . . . This is not a nostalgia for tradition but a call for memory, for a remembering that relations between plants, people, and place have not always been as they are, that they were reorganized through colonialism and continue to be stabilized through" colonial acts, such as Science.[38] This is not how CLEAR works, but it does point to how there are a variety of ways to do anticolonial science without essentialization or appropriation of Indigenous knowledges.

Worldview of Yupiaq Culture"; Knudtson and Suzuki, *Wisdom of the Elders*; and Dene Nation and Assembly of First Nations, "We Have Always Been Here."

Some argue that there is significant overlap between Indigenous and Western science and that they can be integrated, while others work to keep them uniquely separate, even if in collaboration. This is not an argument I will engage with here, except to say that—regardless of the possibility of overlap—academia is rarely an ideal place for Indigenous knowledge, or at least not the academia that I know and work within. It remains hostile to other ways of knowing, except as a source of cultural capital, curiosity, and value for extraction. It remains a Resource relation. This is the context in which I don my lab coat, and it is crucial not to lose sight of that context.

37 As discussed in chapter 1, the emerging drive in academia to capture, incorporate, use, and eat up Traditional Knowledge as a Resource is often another expression of colonialism and the settler and colonial entitlement to Indigenous Land (now with more knowledge!). This trend is why CLEAR *does not* claim to engage in Traditional Knowledge (TK) or Traditional Ecological Knowledge (TEK) collection or use. For more critiques of bringing TK and TEK into the academy and how doing so can reinforce colonial, academic knowledge systems even when that may not be the goal, see McGregor, "Traditional Ecological Knowledge"; Reo, "Importance of Belief Systems in Traditional Ecological Knowledge Initiatives"; Nadasdy, "Politics of TEK"; and Nadasdy, "Anti-Politics of TEK."

For Indigenous readers well versed in these topics looking for a little more nuance, I recommend Duarte et al., "'Of Course, Data Can Never Fully Represent Reality.'"

38 Langwick, "Politics of Habitability," 417, 421.

Maneuvering Knowledge Systems

Navigating this line can be tricky. Of course, Indigenous lab members solve scientific problems in ways that align with traditional teachings and values. For example, after a CLEAR meeting where we discussed how we might discard fish guts in a good way after we had analysed them for plastics, people talked with their fishing families. This is Edward Allen's story:

> *I asked my Elder about "sharing" animal guts. After several moments he shared a memory starting in his childhood. It was my memory as well, and undoubtedly the same memory his Elder kept. When I was young, I was told to take what remains over to feed the dogs, or the birds in the summer months, and these other ones to another place so that the mice might enjoy them. Some were left to be reclaimed by the waters and all that lived below them, and some to go into the ground. As the memory travels through the generations, the only difference was how much there was to take. There was no such thing as waste. All was consumed by us, the animals we shared the land with, or the land itself. Everything is in movement. Even things that were still were gone by morning. Spreading what remains around ensured that they were shared efficiently, and that no remains were piled to the point of contamination. And while the delicacies found in entrails have been forbidden to me because of PCBs and other things from away, the remains still have purpose in the larger whole. They are part of sila and keep me, my Elder, and my Elder's Elder buoyant.*[39]

Edward's conversation with his Elder informed one small part of what is now a regular CLEAR practice: we return fish and other animal guts to the water when our part with them is done. We call this "gut repatriation," but its protocol is not written in the lab book.

Indigenous practices, while they certainly are part of how things happen in CLEAR, are not a shared knowledge system in the lab. As Laurelyn Whitt (unmarked) writes,

> a knowledge system can be defined in terms of four characteristics: epistemology, a theory of knowledge giving an account of what counts as knowledge and how we know what we know; transmission, dealing

39 This story was originally shared in a different version in a lab meeting that is not for public consumption. This written version was prepared for M. Liboiron et al., "Doing Ethics with Cod."

with how knowledge is conveyed or acquired, with how it is learned and taught; power, both external (how knowledge communities relate to other knowledge communities) and internal (how members of a given knowledge community relate to one another); and innovation, how what counts as knowledge may be changed or modified. The systemic nature of knowledge is due to the reciprocal influence of these four characteristics upon one another: how we know, how we learn and teach, how we innovate, and how power figures in this are linked.[40]

It is not that Indigenous sciences constitute one type of thing, and that anticolonial sciences, Eurocentric or Western sciences, and other sciences constitute entirely different sorts of things. Parts of their different knowledge systems overlap.[41] Yet Indigenous sciences have fundamentally different obligations and structures of accountability than other sciences. For instance, CLEAR is not accountable to Edward's Elder, but Edward is, including on the issue of whether and how he shares his Elder's knowledge in the lab. I don't get access to Edward's Elder to ask whether I can share his story in this book: I ask Edward, who asks his Elder. Protocol helps us see our different orientations, different obligations. This is why there is an annoying split in writing out L/land relations in this chapter; some lab members are engaging in good Land relations according to traditional Indigenous teachings, instructions, and obligations, and some are engaging in good land relations as environmentalists, ecologists, ecofeminists, and Nature lovers. Sometimes those obligations overlap, and sometimes they are at odds.

When I think about maneuvering the sometimes overlapping and often-aligned but separate relationships and obligations between Indigenous sciences and dominant sciences in CLEAR (a.k.a. caring), I often think of the two-row wampum. The two-row wampum is a governing document made out of shells (wampum) that illustrates how settlers and Indigenous groups will coexist on separate but parallel paths heading in the same direction.

Of course, the paths are never separate: not in genocide, not in care, and not in anticolonial nor Indigenous sciences. Kim TallBear (Sisseton-Wahpeton Oyate)

40 Whitt, *Science, Colonialism, and Indigenous Peoples*, 31.

41 And not always in a good way. This overlap is a way to describe why Eurocentric sciences find it so easy to appropriate Traditional Knowledge—even if they do not understand that traditional knowledge is a way of knowing rather than what Indigenous people know, scientists can still extract what Indigenous people know, and call it Traditional Knowledge data. But there are also good ways to overlap, such as what Elder Albert Marshall calls Two-Eyed Seeing. See Bartlett, Marshall, and Marshall, "Two-Eyed Seeing and Other Lessons."

writes about how "the difficulties faced by the Native American bioscientists I interviewed cannot be understood within a dichotomy of 'traditional knowledge' versus 'science.' Rather, they can be better understood within a notion of 'harmony' versus the will to know. . . . Almost all travel home periodically, and do not necessarily have trouble reconciling ceremonial practices, or immaterial, 'spiritual' beliefs with the materialist explanations of science. These scientists seem comfortable themselves with having two different knowledge forms at hand to meet their different needs."[42] Here, difference is not the same as mutual exclusion.

These maneuverings work both ways. Not only are Indigenous scientists working to harmonize knowledge systems; so, too, are settler scientists working with aspects of Indigenous protocol. This story is by CLEAR lab manager Kaitlyn Hawkins (settler), who participated in her first gut repatriation ceremony[43] when we were done processing samples. While Indigenous members of CLEAR can burn sage, lay down tobacco, and raise up prayers during the ceremony in Indigenous ways, we also don't expect appropriation of those things. There are other ways to get it right:

> *While touching the guts when returning them to the ocean, we didn't wear gloves out of respect for the animal. I thought at first it would be extremely disgusting, getting blood and guts and some nasty fishy smells all over my hands. Surprisingly though, I didn't mind at all once I started. Don't get me wrong, some of the guts were incredibly messy and extremely smelly, but there was this sort of calm and gratefulness that came over me during the repatriation. I don't recall even smelling the guts (and I know they smelled bad from when I was packing them into coolers earlier in the day). It was just me and the guts and a feeling of peace and gratitude for what the guts had contributed to us.*
>
> *Often while working on guts in the lab, especially when I'm at the lab alone, I'll have a moment where I'll speak to the guts (and the animal that these guts came out of) and do my own kind of appreciation speech for the sacrifice that these animals had made for us (as both a source of food and for science). It felt very fitting that I was then a part of the repatriation, almost as if I were saying my goodbyes to these animals that I had grown intimate with (in a sense) from my work in the lab.*

42 TallBear, "Indigenous Bioscientists," 183.

43 Some (usually Indigenous) audiences have asked about the ceremonialization of this protocol. This protocol does ceremony more in the way that taking a chocolate bar out of your back pocket is a prayer than something more formal. Rick Chavolla (Kumeyaay), CLEAR's (and my) Elder advisor, helped us figure out how to return guts to the Land the first time we did it.

To be honest, I was nervous and wasn't sure what to expect before the repatriation. I wasn't sure that I really knew what I was supposed to be doing and how I was supposed to act. Those feelings quickly passed once we began, and I just sort of understood what it was that I was supposed to be doing and how to act. It's hard to describe in words how I felt, but I felt at peace with returning these guts back to the land. I felt like we were doing the right thing by honouring them this way rather than just tossing them in the trash like a man from the wharf had suggested when he first saw us putting the guts in the water. It felt respectful, like it was the right and the best thing that we could do for the animals. It left me with an incredible sense of calm.

Once we had finished returning our guts to the land, we took a quick hike up the Sugarloaf Trail so we could get away from some of the hustle and bustle of the wharf and have a quiet moment. While standing on a cliff and looking out over the harbour and the ocean, we saw several humpback whales swimming in the distance. This was the perfect ending to the repatriation, showing us the circle of life and how everything in the ocean is connected. It further added to my feeling of calm and strengthened that we had done the right thing by returning the guts to the land. I am so glad I was able to experience and be a part of the repatriation!

One of the things that makes Kaitlyn Hawkins an extraordinary lab manager for CLEAR is her ability to make space for others, from whatever position they are starting from. In this quote and in many other ways, she exemplifies what Kim TallBear calls "standing with," where knowledge is co-constituted "in concert with the acts and claims of those who I inquire among,"[44] whether that is other lab members or fishes. Thank you, Kaitlyn.

Anticolonial Sciences

Anticolonial sciences, even when they run parallel to or overlap with Indigenous sciences and practices, make space for settler and other scientists as well as allies in unexpected places. The university's protocol for disposing of animal tissue, like fish guts, is to incinerate them as biohazardous waste. CLEAR had to get permission to deviate from this regulated practice. When I emailed the biosafety group at the university, I was ready to fight—I was requesting something counter to policy, to practice, even to regulation. They emailed back a one-sentence reply that said to go ahead, repatriate those guts. I was so surprised I nearly forgot to think of them as allies. Finding allies in unexpected places,

44 TallBear, "Standing with and Speaking as Faith," 5.

recognizing the many ways that "colonial schools become disloyal to colonialism,"[45] and understanding that power is not a monolithic wall to throw your soft body against are all important parts of an anticolonial science.

As is the case for Indigenous sciences, there are many different types of anticolonial sciences, and there are overlaps between anticolonial sciences and what we call Science(s), colonialism(s), resistance(s), and L/land.[46] None are monolithic or stable, but rather changing, moving, patchy, incomplete, plural, and diverse.[47] Often I hear scholars and activists alike talking as if capitalism (or patriarchy or racism, but mostly capitalism) is a solid monolith that we must dash our soft bodies against, to little avail. But that characterization gives capitalism and colonialism more power than they merit by erasing not only their diversity, but also the patchiness, the unevenness, and the failures of those systems to fully reproduce themselves.[48] It erases the other kinds of economies and L/land relations that happen within, alongside, and in spite of capitalism, the university, and colonialism. So let's not.

Even within dominant science, there are many anticolonial sciences: queer science,[49] abolitionist science,[50] Zapatista science,[51] feminist science,[52] anarchist science,[53] slow science,[54] anticapitalist and communitarian science,[55] and sci-

45 paperson, *A Third University Is Possible*, xvi.

46 The refusal to separate out who is in a land relation and who is in a Land relation, especially as these relations shift and overlap, is why the annoying L/land will appear for the rest of this chapter. It's ugly but it's truer.

47 For a similar discussion of capitalism, and how giving capitalism a monolithic architecture misses already-existing resistances against it, see Gibson-Graham, "End of Capitalism (as We Knew It)"; Peck, "Explaining (with) Neoliberalism"; and Neville and Coulthard, "Transformative Water Relations."

48 Thank you, Josh Lepawsky (settler), for your input about this problem in academic conversations. This also builds on the work of J. K. Gibson-Graham and la paperson. To extend this, if capitalism, the university, etc. are hard monoliths and all we can do is dash our bodies upon them, then the only form of activism is to become a bloody body. There is a greater diversity of activism. David and Goliath is a stupid model for change.

49 E.g., Mortimer-Sandilands and Erickson, *Queer Ecologies*; van Anders, "Van Anders Lab."

50 E.g., Rusert, *Fugitive Science*.

51 Duncan, "Zapatistas Reimagine Science as Tool of Resistance."

52 E.g., van Anders, "Van Anders Lab."

53 E.g., Thorpe and Welsh, "Beyond Primitivism."

54 E.g., Stengers, "Another Science."

55 E.g., Bencze and Alsop, "Anti-Capitalist/Pro-Communitarian Science and Technology Education."

ences from below,[56] among others.[57] So why not just say we're doing intersectional feminist and queer science, then? First, queer, feminist, and other sciences are not monolithic or stable either—some expressions of these sciences can be colonial in their entitlement to Land.[58] Foregrounding colonialism avoids the implication that queer or feminist orientations and obligations are automatically and simultaneously anticolonial orientations and obligations. An anticolonial science does not conflate and collapse different forms of oppression and resistance into one category.[59]

This specificity is also why we do not say CLEAR does decolonial science. Eve Tuck (Unangax) and K. Wayne Yang (diasporic settler of colour) write, "Decolonization, which we assert is a distinct project from other civil and human rights–based social justice projects, is far too often subsumed into the directives of these projects [talking about social justice, critical methodologies, or approaches which decentre settler perspectives], with no regard for how decolonization wants something different than those forms of justice."[60] Unlike anticolonialism, which can take many forms, "decolonization specifically requires the repatriation of Indigenous land and life. Decolonization is not a metonym

56 Harding, *Sciences from Below*.

57 There are many labs out there doing research with an orientation to good land relations. Here are just a few: the Technoscience Research Unit, Political Conocimiento Development Tools Lab, The Data Warriors Lab (forthcoming), Collaboratory for Indigenous Data Governance, Forensic Architecture Organization, Apology Lab, Global Witness Lab, the Mother's Radiation Lab—Tarachine, Dine No Nukes, Sitka Tribe of Alaska Environmental Research Lab, Indigenous Community Based–Health Research Lab: Morning Star Lodge, Te Koronga, Indigenous Futures, Fab Lab Palestine, Hyphen Labs: NeuroSpeculative AfroFeminism, Decolonial Sustainability Lab, UHURU Black Liberation Lab, Indigenous STS–Indigenous Peoples in Genomics Canada (SING Canada), INDIGI LABS, DREC (Digital Research Ethics Collaboratory), Ida Wells Lab, Onman Collective, Black Farmers Collective, Corp Watch, Black Mesa Water Coalition, Ngā Pae o te Māramatanga New Zealand's Māori Centre of Research Excellence (CoRE), Liberation Lab, and Feminist Approach to Technology (FAT) lab . . . among so, so many others. If you run a lab that works from anticolonial and other good land relations, please introduce yourself to me!

58 See the introduction for discussions of environmentalist and anticapitalist alignments with colonialism.

59 This is a key argument in Tuck and Yang, "Decolonization Is Not a Metaphor." I *highly* recommend reading it if you've ever claimed to decolonize something, or even if you are just attracted to the idea. Another good nuance is the second part of chapter 1 in Tiffany Lethabo King's *The Black Shoals*.

60 Tuck and Yang, "Decolonization Is Not a Metaphor," 2, 3.

for social justice,"[61] though I also acknowledge that different colonialisms will have different decolonialisms and anticolonialisms. Does CLEAR do decolonial work? Do we repatriate Land and Life? Sometimes/perhaps, but I don't think that is for public consumption. By my understanding, most of what CLEAR does is not decolonial.

Anticolonial sciences are characterized by *how* they do not reproduce settler and colonial entitlement to Land and Indigenous cultures, concepts, knowledges, and life.[62] They are not "about rescuing settler normalcy, about rescuing a settler future."[63] There are a lot of ways to do anticolonial science. For example, in "Being a Scientist Means Taking Sides," biologist Mary O'Brien (unmarked) contends, "There are infinite questions that you could ask about the universe, but as only one scientist, you must necessarily choose to ask only certain questions. Asking certain questions means not asking other questions, and this decision has implications for society, for the environment, and for the future. The decision to ask any question, therefore, is necessarily a value-laden, social, political decision as well as a scientific decision."[64] While this argument is ideal for teaching people about how values are inherent in the supposedly neutral, objective, and culture-free domain of dominant science, I found O'Brien's work while seeking scientific critiques of assimilative capacity (see chapter 1). As a professional biologist, she writes against the ubiquity of risk assessment as a dominant scientific framework to describe harm. In so doing, O'Brien implicitly argues against an entitlement to Land as a sink for pollution:

> By diligently preparing and analyzing data for risk assessments . . . scientists are participating in the process of assimilative capacity assessments and policymaking rather than alternatives assessments. Assimilative capacity assessments ask, How much dioxin is safe in the milk of an infant's mother? How much hazardous waste can be burned without raising the cancer risk to nearby residents by more than one in a million, or one in a hundred thousand, or perhaps one in ten thousand? . . . One could ask instead, What alternatives do we have to the industrial use of chlorine,

61 Tuck and Yang, "Decolonization Is Not a Metaphor," 21.

62 I used to say "lifeworlds" but I really mean bare life, as Tiffany Lethabo King reminds us: "Genocide—and the making of the Native body as less than human, or flesh—remains the focus and distinguishing feature of settler colonialism that is worth defining and analytically parsing for readers." King, *Black Shoals*, 56. Repatriate Land and *Life*.

63 Tuck and Yang, "Decolonization Is Not a Metaphor," 35.

64 O'Brien, "Being a Scientist Means Taking Sides," 706. This is an excellent teaching text if you are working with young scientists and engineers.

which results in the placement of dioxin in an unborn embryo's tissues? What alternatives are available to reduce toxics use and generation of hazardous wastes and eliminate the making of cement by burning solvents and other toxics? What social and production alternatives do we have to cutting the last of our ancient forests? What is the least habitat we can take away from a species in trouble? . . . I contend that, in general, to ask risk-assessment questions rather than alternatives-assessment questions is to contribute to the currently dominant, but suicidal, assimilative capacity approach and practices of our society. Many industry associations adopt the assimilative-capacity approach, because the questions asked support extractive and polluting activities.[65]

O'Brien does not have to talk about Land, Indigenous peoples, or even justice to practice dominant science with anticolonial elements. As discussed earlier, "to be subject to anti-Indian technologies does not require you to be an Indigenous person."[66] Colonialism and its pollution affect a wide range of peoples. So, too, can a wide range of peoples engage with anticolonial technologies like the ones O'Brien discusses.

To be clear, anticolonial sciences are not just technical tweaks to dominant science. Anticolonial sciences function more like infrastructures, underlying "the ways knowledge-making can install material supports into the world—such as buildings, bureaucracies, standards, forms, [instruments], funding flows, affective orientations, and power relations."[67] I am proposing anticolonial sciences as knowledge systems, sometimes arrayed with, sometimes adjacent to, and sometimes explicitly against the knowledge systems of dominant science.

Because knowledge systems are based on reciprocal influences of how we know, what we learn and teach, how we change, and how power works, CLEAR does not operate by tweaking protocols (though we do that, too). We don't add a bit of land theory here, and work to be a little less elitist over there. Instead, we aim to transform every moment of every aspect of our research, from how we pay people, to sampling methods, to peer review, into good relations with L/land and against dominant scientific relations with L/land based in separation, universalism, maximum use, unfettered access, standing reserve, and proofs of harm (among other things). Leonie Pihama (Māori) reinforces this way of being and doing, linked to research and everyday life:

65 O'Brien, "Being a Scientist Means Taking Sides," 706.
66 paperson, *A Third University Is Possible*, 11.
67 Murphy, *Economization of Life*, 6.

> It seems that every day I get a request to meet or talk with organisations about how to "do" Kaupapa Māori Research Methodology. Starting point is that you don't "do" Kaupapa Māori. You live Kaupapa Māori. You live tikanga. You live te reo. You live as fully Māori as you can. You strive for rangatiratanga for our people. You seek to transform injustice and colonial oppression in all its forms. You live aroha, manaakitanga, kaitiakitanga. You honour mana, tapu, noa. You uplift & affirm mana wahine, mana tane, mana Tangata, mana atua, mana Whenua. Then we can start talking about what Kaupapa Māori Theory and Methodology really looks like.[68]

Compromise and Obligation

Compromise

Regardless of the specifics of your approach, doing anticolonial science within a dominant scientific context is simultaneously a commitment to dominant science and a divestment from it, which makes it uniquely compromised. Compromise is not about being caught with your pants down, and it is not a mistake or a failure—it is the condition for activism in a fucked-up field. Research and activism, scientific or otherwise, never happen on a blank slate. As a result, we are always caught up in the contradictions, injustices, and structures that already exist, that we have already identified as violent and in need of change.

During her activist research in Bhopal, India, after one of the largest industrial chemical disasters in history, STS scholar Kim Fortun (unmarked) reflects: "Idealized portraits of advocacy represent a certainty that is resolutely at odds with how environmental problems materialize on the ground, in continuing negotiations over what is real, what is past, and what is to come. Described in ideal terms, the advocate is never seen enmeshed in discrepancies, ambiguities,

68 Pihama, "It Seems That Every Day . . ." Are anticolonial sciences knowledge systems or methodologies? I don't really care. They are a commitment and an orientation, and they play out at multiple scales simultaneously, which means they do not focus on the scale of technical tweaks like adding more Indigenous people to the lab, to the readings, or as data collectors (since inclusion or diversifying empire does not change anything about structures, infrastructures, power relations, or L/land relations). Knowledge systems and methodologies are understood in different ways by different disciplines and groups, but they both look at scales that cover epistemology rather than just knowledge, ethics instead of just ethics review forms, and assumptions rather than truths.

and paradox. Nor is he seen trying to force fit the world into available political ideologies.... Idioms for ethics without full knowledge remain undeveloped."[69]

Compromise means maneuvering the "discrepancies, ambiguities, and paradox" of doing anticolonial science in a dominant scientific field, "trying to force fit the world into available political ideologies." Charles Hale (settler) argues that any act of resistance is "partly implicated in the very systems of oppression they set out to oppose,"[70] inadvertently reproducing parts of the system while challenging or changing others. There is no recourse to purity, where anticolonial scientific activism and resistance is outside of, free of, or separate from the colonial systems we seek to oppose.

In chapter 1, I mentioned that CLEAR no longer uses chemicals that require hazardous waste disposal, because hazardous waste disposal assumes access to Land as a sink. This restriction includes not using KOH, which in turn limits our ability to study bivalves, crustaceans, and other invertebrates for plastic ingestion since you need KOH to dissolve their shells. But recently, Liz Pijogge (Inuk), my main research partner in Nunatsiavut, said that she wants us to investigate mussels and whelks for plastics because they are key parts of traditional food webs. The compass that allows us to pick through this conflicted terrain is a commitment to good L/land relations and a commitment against reproducing colonialism. We don't have an answer yet, but we have a way forward.

It's a lot of work. In *Research Is Ceremony*, Shawn Wilson (Cree) writes:

> Like myself, other Indigenous scholars have in the past tried to use the dominant research paradigms. We have tried to adapt dominant system research tools by including our perspective into their views. We have tried to include our cultures, traditional protocols and practices into the research process through adapting and adopting suitable methods. The problem with that is that we can never really remove the tools from their underlying beliefs. Since these beliefs are not always compatible with our own, we will always face problems in trying to adapt dominant system tools to our use.[71]

69 Fortun, *Advocacy after Bhopal*, 52. Thank you, Kim Fortun, for the work you have done and continue to do. Your willingness to share ideas and ask questions with others, to learn from others (including your juniors), to stand up to others (including your seniors) has impressed me during our interactions. Thank you.

70 Hale, "Activist Research v. Cultural Critique," 98.

71 Wilson, *Research Is Ceremony*, 13. Also see Lorde, *Master's Tools*.

One response to this incommensurability is to move to Indigenous science, with its own knowledge systems. But what if you are rooted in and committed to dominant research paradigms?

Zoe Todd's (Métis) work is often about this in-between space (also about fish). Her work, as well as that of many other Indigenous scholars in academia, "attend[s] to the space between and across a) the Euro-Western legal-ethical paradigms that build and maintain the academy-as-fort (or colonial outpost), fixing it within imaginaries of land as property and data as financial/intellectual transaction, and b) Indigenous legal orders and philosophies which enmesh us in *living* and *ongoing* relationships to one another, to land, to the more-than-human, and which fundamentally challenge the authority of Euro-Western academies which operate within unceded, unsurrendered and *sentient* lands and Indigenous territories in North America."[72] The question is (and for many of us, always is): how do we do our research in this space between?[73] The answer will play out differently for settlers, and for different kinds of settlers and non-Indigenous people that aren't settlers, than for Indigenous scientists (and different kinds of Indigenous scientists!).

Obligation

Compromise is what happens when you have obligations to incommensurabilities. Incommensurability means things do not share a common ground for judgment or comparison; that is, "projects that simply cannot speak to one another, cannot be aligned or allied."[74] Anticolonialism within dominant science. Diversity work in a racist institution.[75] Humility in a tenure application. All are impossible bedfellows that are nonetheless crucial to pursue and indeed happen, yet should never be smoothed over or conflated in that pursuit. Tuck and Yang write, "An ethic of incommensurability . . . recognizes what is distinct"[76]

72 Todd, "From Fish Lives to Fish Law"; emphasis in original. A note on unceded land: Loretta Ross (Bimaashi Migizi), treaty commissioner of Manitoba, has explained that even treaty Land is unceded, for the Land was never available to be ceded to begin with. To give up ownership of Land is a colonial possibility, not one rooted in most Indigenous concepts of Land. Ross, "Treaties, Reconciliation, and Me."

73 This is not a new question. See, e.g., Fine, "Working the Hyphens"; Donald, "Indigenous Métissage"; Wilson, *Research Is Ceremony*; paperson, "Ghetto Land Pedagogy"; paperson, *A Third University Is Possible*; Moten and Harney, *Undercommons*; and Todd, "Refracting Colonialism in Canada."

74 Tuck and Yang, "Decolonization Is Not a Metaphor," 28.

75 Ahmed, *On Being Included*.

76 Tuck and Yang, "Decolonization Is Not a Metaphor," 28.

and what cannot be joined even if allied. In an anticolonial science, for example, "an ethic of incommensurability means relinquishing settler futurity, abandoning the hope that settlers may one day be commensurable to Native peoples."[77] At the same time, "We argue that the opportunities for solidarity lie in what is incommensurable rather than what is common across these efforts."[78] An ethics of working with and through incommensurable values, futures, and knowledge systems "brings these areas into conversation, without papering over the differences, but also without maintaining false dichotomies."[79]

I understand a commitment to anticolonial science as one rooted in incommensurabilities that nevertheless moves forward with, in, and around impossible bedfellows. Rather than erasing or smoothing difference, or claiming that something is incomprehensible because it does not align with what makes sense on my/our/your logics, or reaching resolution or consensus, I understand an ethic of incommensurability as one that digs into difference and maintains that difference while also trying to stay in good relations.

Admittedly, an ethic of incommensurability within anticolonial science is hard to wrap my head around. But only in theory. On the ground, it is easier because my obligations are clear. In the last chapter, I talked about how scale is a way to describe which relationships matter within a given context (e.g., gravity and capillary action exist for both elephants and viruses, but gravity matters more to elephants and governs their movements, while capillary action matters more to viruses and governs their movements). For obligation in the anticolonial science that CLEAR engages in (and I would guess for most forms of obligation), the relationships that matter are not between yourself and the system, but with the collective and systems. Obligations do not exist and are not enacted in atomized and individualized one-on-one relationships but in a diversity of relationships where some relations matter more than others.

I'm here because I'm a scientist and I need fish guts, plus I love fishing. I'm the guest of my friend and two men I've just met from Nain, the most northern settled town in Labrador, Canada. It's my first time fishing for Arctic char. I'm excited. But when I look around the boat, there is no fish bonker—that wooden stick you use to bonk the fish over the head. Uh oh. How do you kill the fish? I throw out my line so I don't catch anyone. The men are catching char after char, throwing them over their shoulders into the fish boxes and

77 Tuck and Yang, "Decolonization Is Not a Metaphor," 36.
78 Tuck and Yang, "Decolonization Is Not a Metaphor," 28.
79 Tuck and McKenzie, *Place in Research*, 5.

casting out again without looking back. The fish suffocate. When my friend catches her first fish I ask if I can kill it. She nods. I grip the fish between my knees and I cut its throat, sawing and sawing away. My borrowed knife is dull. The skin is unexpectedly thick. Shit, shit, shit. I'm doing a bad job. The char is bleeding and bleeding all over me, but is still alive. I cut deeper and deeper, past its gills. Still alive. Shit. Past its eyeballs. Its head is nearly off. Still alive. I am covered in blood and flies. I throw the char in the fish bucket, where it bleeds out. Shit. The next time my friend catches a fish I cut deeper and faster. Still not good. She lands a third fish. I feel sick. I am going to have to keep sawing away like a shitty idiot. The boat driver interrupts his casting to pick the fish out of my hands by its tail and whack its head against the side of the boat. Dead. Thank god. I start catching fish and whacking them. Though only mine and my friend's. Turns out you don't slit a char's throat. You have to whack char. I really need to add that to the lab protocol.[80]

The way people in the boat were catching and killing char came from their relations. My (strained, evolving) methods of killing fish by slitting throats and whacking heads came from my obligations to my father, who taught me to catch and kill fish quickly; to my Elder, who taught me about good relations in general (including keeping a sharp knife); to my friend on the boat, who invited me to her home in Nain; to the men on the boat, who invited me onto the boat; and to the fish, who died. I killed my and my friend's fish, but not the men's. That was not my place. One of the men, who was accountable to his guest, showed me how to bonk the fish. The men had other obligations, too, that made them throw the char in the box as fast as they could—they had to fill freezers (theirs and other people's) in a place with acute food insecurity. Perhaps. I was a guest, so this is speculation and none of my business. The point is: different relations make different obligations, which engender different methods. This is not relativism, but a deep specificity based in place and in the relations to which we are accountable.

Community Peer Review

The first time I found plastic in a cod, I was thrilled. I remember seeing the small white plastic on my blue gloved finger, the elation of finally finding a fragment of plastic after looking through nearly a hundred stinky cod guts. Then I suddenly realized what I had done. I had found plastics in *cod*. In *Newfoundland*. Shit.

80 This story also appears in M. Liboiron et al., "Doing Ethics with Cod."

Since the sixteenth century, when colonial fisheries began in earnest on the island of Newfoundland, cod have shaped the socioeconomic structure of the colony and then the province, the colonial geography of where people live, and settlers' ability to sustain themselves.[81] In the 1970s, "the world's best funded, most prestigious, scientific fisheries management system"[82] managed the wild cod populations—a necessity as the province started to prioritize industrial-scale fishing. Spoiler: a change in scale from sustenance to industrial fishing is a change in relations that matter. The government scientists missed this. On July 2, 1992, the settler government declared the cod fishery had collapsed and called a moratorium on cod fishing. An estimated 9,000 to 10,000 fish harvesters and somewhere between 10,000 and 12,400 fish-plant workers lost their income in a province of 580,109 people.[83]

The cod moratorium criminalized sustenance fishing, which was central to ways of life and living in the province. When I walked into my first rural (here: "outport") restaurant in 2014, the only thing hanging on the wall was the newspaper article announcing the 1992 cod moratorium. When we talk about the moratorium in undergraduate classrooms, students who weren't even born in 1992 talk about it as if it's fresh in their memories. When I teach statistics and we're using temporal zones for analysis, students say "1992" is the sharpest temporal marker for any analysis of the island no matter what we are analysing. I don't think I've ever taught a student from the island whose family wasn't directly affected by the cod collapse. Cod matters here, and fisheries science killed cod.

And I just found pollution in a cod, as a scientist.

I remember staring at the little plastic fragment on my finger and thinking, "*How* am I going to handle this?!" What are my obligations? *How* do I not cause harm? Then I thought, "How would I know? I have to ask Newfoundlanders." CLEAR's community peer review process was born in that moment.

If colonialism is a mode of domination where settlers and colonial forces have access to Land for their goals, including the conduct of environmental research, then community peer review is a way to cockblock that entitlement.[84] Like traditional academic peer review, community peer review is a way to ad-

81 Ommer, "After the Moratorium."

82 Bavington, *Managed Annihilation*, 13.

83 Gien, "Land and Sea Connection"; Schrank and Roy, "Newfoundland Fishery and Economy."

84 Don't worry, I looked up *cockblock*, and it really is all one word. Makes sense not to have any extra spaces where things might wiggle through.

FIGURE 3.3. Community members looking at plastic samples as part of community peer review. Photo by Bojan Fürst (settler) from the photographic series *How We Do Science* (2018).

judicate and distribute knowledge that enacts the values of community self-evaluation, ethical procedure, and reliability of results. It is a way to meet obligation, with others, in context. But CLEAR's protocol for community peer review outlines a markedly different view of peers, ethical distribution, quality, and reliability than that of the academy. Yet the procedure of community peer review is quite similar to academic peer review: researchers share work with a community of peers (in this case, local fish harvesters and community members), discuss what they did, why they did it, and what they found, and the reviewers give input to make the work better and either deny or support its publication.[85]

For CLEAR, community peer review is about obligations to Land, including fish, fish harvesters, local ways of knowing, events of the cod moratorium, and more. The colonial assumption of many research practices is that researchers have a right to access Land for data acquisition. But researchers are not entitled to conduct research on someone else's L/land, whether it falls under private title or collective land claims or is part of homelands. Land is always part of a community, whether there are humans present or not. Feminist geographer

85 For a step-by-step, protocol summary of community peer review, see M. Liboiron, Zahara, and Schoot, "Community Peer Review." This document is a preprint and has not been published at the time of this writing.

Doreen Massey (unmarked) has critiqued "a persistent identification of place with 'community.' . . . What gives a place its specificity is not some long internalized history but the fact that it is constructed out of a particular constellation of social relations, meeting and weaving together at a particular locus."[86] We can extend this concept of community to include people who aren't human, materials, landscapes, events, obligations, and other types of relations.

One of the first steps in CLEAR's protocol for doing community peer review is to do your homework to "understand the wider historical and political context of the community."[87] While this can be met in part by "reading texts by Newfoundlanders about Newfoundland, reading local newspapers and Fishermen's Unions' annual reports," it is more important to hire "local graduate students and field technicians [from the community] to be part of the process."[88] Obligations to Land and community without locals are weird and unlikely, even impossible. For CLEAR, homework has helped us determine whether and which fish are part of the community,[89] as well as the legacies of fisheries science and food sovereignty that characterize CLEAR's research contexts and obligations. It also *must* precede any act of "outreach" to speak with community members, since having locals on the research team gets rid of the creepy "out" in "outreach,"[90] as it starts to blur (but never gets rid of!) the lines between the research community and the fishing community, and brings in more robust forms of accountability.

CLEAR's community peer review protocol does a variety of work, such as creating a way to recognize more forms of harm and violence beyond those usually thought about by science professionals or captured in scientific research designs (see chapter 2); making space for narratives about fish, food, and pollution beyond deficit models and damage narratives;[91] making space for guidance, analysis, and collaboration from experts outside of academia; and, perhaps most importantly for anticolonial science, setting the stage so that communities can refuse our research.

86 Massey, "Global Sense of Place."

87 M. Liboiron, Zahara, and Schoot, "Community Peer Review," 9.

88 M. Liboiron, Zahara, and Schoot, "Community Peer Review," 9.

89 Todd, "Fish, Kin and Hope." CLEAR's community peer review protocol is currently anthropocentric and is built for humans. We are working on a protocol that would also include others. Fish and others can communicate and have the ability to refuse, so this method should work more broadly. Stay tuned. See Wadiwel, "Do Fish Resist?"

90 If you've ever been the subject of outreach, you know what I'm talking about. All reachy and getting into your business.

91 Tuck, "Suspending Damage."

The term *refusal* in a research context refers to ethical and methodological considerations about how and whether findings should be shared with and within academia at all, as researchers often encounter information that might be intensely personal, fundamentally contextual, sacred, intended only for certain people, or needs to be earned.[92] Offering refusal is part of ensuring research is accountable to its relations. In refusal, rather than "the terms of accommodation . . . being determined by and in the interests of the hegemonic [more powerful] partner in the relationship"[93] such as academics, fish harvesters and villagers set the terms of how and whether research that impacts their communities should occur, be conducted, and circulate. Practicing refusal keeps community knowledge from being a Resource.

As such, refusal is affirmation and repair more than denial (though it's certainly that, too!). Refusal "is not just a 'no,' but a redirection to ideas otherwise unacknowledged or unquestioned."[94] It can highlight and address the strained relationships between academics and communities,[95] realign research values to local needs, benefits, and protocols, and, of course, bring attention to how the right to research is a colonial concept.[96]

We've been refused before. When CLEAR was conducting community peer review in a small fishing village, someone in attendance asked if we worked with the Department of Fisheries and Oceans (DFO), the federal arm tasked with regulating fisheries. It also happens to be the source of the fisheries science that

92 There is a strong and growing literature on refusal in Indigenous thought that articulates refusal as a methodology, an ethic, a politics, and a right. This work includes but is not limited to Gaztambide-Fernández, "Elite Entanglements"; Coulthard, *Red Skin, White Masks*; Grande, *Red Pedagogy*; Grande, "Refusing the University"; McGranahan, "Theorizing Refusal"; Moffitt, Chetwynd, and Todd, "Interrupting the Northern Research Industry"; A. Simpson, "On Ethnographic Refusal"; Zahara, "Refusal as Research Method in Discard Studies"; A. Simpson, "Consent's Revenge"; Tuck and Yang, "Unbecoming Claims"; Tuck and Yang, "R-Words"; Wadiwel, "Do Fish Resist?"; and A. Simpson, "Revenge of Consent."

93 Coulthard, *Red Skin, White Masks*, 17.

94 Tuck and McKenzie, *Place in Research*, 147.

95 Smith, *Decolonizing Methodologies*.

96 When I've presented the methodology of community peer review to mostly white settler audiences, especially those that work with Indigenous groups, people are discursively on board, but questions tend to stray toward techniques of ensuring research is not refused. That's creepy (see chapter 2 for more on creepiness as settler desire). Giving up the entitlement and perceived right to data is a central—the central!—ethic of anticolonial sciences. You will lose things in anticolonial research that you automatically get as a researcher in colonial modes of research (regardless of your heritage).

led to the 1992 cod collapse. It was clear that if we worked with DFO, we were not welcome.

This example is one of the rare explicit refusals we have encountered in community peer review. For the most part, people are good hosts. They ask us how we are doing and how our families are. They smile and say "welcome." It can be easy to confuse good hosting for consent. One of the crucial aspects of community peer review is that, like consensual sex, refusal can be indicated by something other than a clear "no." We have watched our colleagues' informants welcome them into their communities, feed them, give them places to stay, and then refuse at every other stage of the research by not showing up to interviews, coming late, saying questions or tasks are too complex, saying they don't know the answer to obvious questions, or telling researchers they should talk to so-and-so, who is unavailable because they are on the Land for the next three months, the local sell-out, or dead.[97]

It can be hard to see your obligations, especially when they are counter to your desires. But there are ways to help, with the collective, in context. Part of CLEAR's community peer review protocol is to have several note takers write observational notes during community review meetings (with consent!), writing what attendees did, what they said, how many there were, where they sat, what body language they displayed and if/when this changed, among other observations. These are our clues and cues to refusal. Then we read those cues and clues. Analysis is tricky. During an academic presentation, it's considered quite rude for someone to answer their cell phone. But in some of the small villages we've been in, it's normal. What if a community member says the work sounds important, and we should go do it in a neighbouring village? Is that an invitation to leave, or to extend the research? What if no fish harvesters show up? Is it because you are being refused, or that you didn't do your homework properly and there's a hockey game on that evening? (That's happened.) If you bring fancy tea biscuits to the meeting and no one eats them, are you being refused, or do local tastes run more toward Tim Horton's doughnuts? (That's happened.)

The answer is: if you are not from that place, you don't know.

This is why it is crucial to community peer review to hire local people as full researchers on your team, from the start and with a place in decision making and analysis.[98] There is no other way to do this method. No amount of reading,

97 For more on what refusal can look like, see A. Simpson, "On Ethnographic Refusal," 77.

98 The ideal here would be to hire so many, so often, that the community can do its own research and you make yourself obsolete as an outsider. That's our goal. That way, communities have a real choice to work with us, or not.

asking around, or observing will tell you. This lab story is from Saskatchewanite CLEAR member Alex Zahara (settler):

> *I was sitting in the back of a truck, surrounded by lab members as we returned to St. John's after our lab's first official public meeting. The meeting was held nearly twenty minutes away in a nearby fishing community, and as a PhD student at Memorial for just under three weeks, the late January evening was one of my first times out of town to visit rural Newfoundland. As the city lights grew bigger, I remember noticing how the snow contrasted so differently against the scattered grey of frozen ponds and the blueish black ocean waves, which I could see whenever we drove up a hill. Like many of the other CLEAR lab members, I'm a come from away (the local term used to describe non-Newfoundlanders) and the landscape was so different from the Canadian Prairies that I was used to.*
>
> *Lab members explained to me that the public meeting was meant as a way of being accountable to the communities involved in our cod fish study. In practice, I observed that this meant taking on different roles: Max, the lead of the study, presented our findings and responded to most of the audience feedback; Charlie and Emily, who participated in data collection and analysis, stood near displays of sample plastics; and still others, such as myself, handed out surveys, took notes about demographics, recorded audience questions, mood, and responses. As a new lab member, writing notes at the meeting was important for learning about our audience (were they anti-plastic activists or local fishers? maybe overlap between the two?) and to better understand their concerns. As we sat together in the truck, we debriefed on what we learned through our respective tasks.*
>
> *Being a science lab, perhaps it wasn't surprising that many of our initial observations were quantitative: there were more women than men attending the meeting (surely this was a feminist observation?[99]), and the group was a near equal mix of both older and younger people. But as we continued driving along, we began pointing out more subtle happenings: Did you also notice that nobody sat in the front row? How about that people's arms were crossed? Did anyone else write down that people were really quiet at the beginning? These moments, however, often contrasted with what people said to us directly: people indicated support of our research findings, and also gave us suggestions for future research. As we approached campus, we quickly brainstormed reasons for this disconnect. And looking back, I think my top-of-the-*

99 LOL.

brain suggestions weren't totally helpful (maybe Newfoundlanders are just shy? Do they not typically like marine scientists?), though not knowing the place, perhaps this was to be expected.

At the time of this writing, our paper on how we conduct community peer review has been rejected by an open-source science journal that has a special section on peer review. The stated reason is that the method is not universally replicable. We're glad they understood the method! Peer review will always change based on the communities of peers doing the reviewing. The rejection itself is the proof.

On the bright side, several of CLEAR's scientific papers have passed both community and academic peer review and have been published. When I give presentations on community peer review, people often ask what would happen if publication were refused. The short answer is: We would follow the instructions of the community. We wouldn't publish. Perhaps the community thinks the knowledge is more properly held by the local fishermen's union than by a scholarly publication. Then that's where we put it. Research does not have to be published to be valid. This also means that community-situated research is risky for graduate students whose thesis might be refused. This method underlines how community-based research is not inherently lovely. It requires ethics that can cause loss, rather than only gain, for researchers. It must be so: otherwise it's a Resource relation.

So far, all of CLEAR's findings that have been approved and improved through community peer review have also been approved for publication (though if they weren't, would I be able to tell you?). We've had parts refused, been directed to new and different research questions, been told some of our analyses are wrong, and one group asked us to tell you: Atlantic cod tested from the island of Newfoundland have a lower plastic ingestion rate than cod tested in Norway.[100] All of these insights and requests have been honoured.

100 Apparently, Norway is one of Newfoundland's main competitors for Atlantic cod. This is the main metric that mattered during one community peer-review meeting. We now always include comparisons of the same commercial species in nearby places in our papers. Nuances like this, rather than total refusals, characterize most of our community peer-review feedback.

Place-Based Sciences

Community peer review is place-based, where "the relationship between an object and where it belongs is not simply fortuitous, or a matter of causal forces, but it is rather intrinsic or internal, a matter of what that thing actually is."[101] Place-based science is not unique to CLEAR. Historian Geert Somsen (unmarked) has written about a broader international movement where "the current emphasis on local contexts is not only opposed to a European picture, but also to the long-standing notion of science as inherently universal. In fact, the localist perspective has developed precisely as a reaction against such universalism: the idea that science is independent of the place where it is practiced (because of the universality of its knowledge or method), and that scientific practitioners are automatically united in a single global pursuit."[102]

An argument for the emplacement of knowledge, including scientific knowledge, has long been made in feminist science studies. One version of this argument is that it is impossible to distinguish an object of study from such tools of observation as microscopes and sensors that collect some data and not others.[103] The standpoint one is observing from, including one's own body, social location, history, and training, among other situating and emplacing factors, also shapes what one sees.[104] These insights about the emplacement of knowledge extend to obligation. If we make the things we observe, then, as Angela Willey (unmarked) asserts, "we become responsible not only for the knowledge we seek but for what exists."[105] This is not to say that place-based knowledge is inherently good. Indeed, colonialism is also a place-based practice. This is to say that place engenders a specific type of relationship with its own set of compromises and obligations that do not assume that universal laws govern L/land relations.

Judgmental Sampling

Some science speak: to count scientifically, you first must organize the world[106] so you can be sure that your count represents the thing you are trying to study. Sampling is one way to do this. Since you usually can't count every single instance

101 Curry, *Digital Places*, 48.

102 Somsen, "History of Universalism," 362.

103 E.g., Barad, *Meeting the Universe Halfway*; Murphy, *Sick Building Syndrome*.

104 E.g., Haraway, "Situated Knowledges"; Harding, *Feminist Standpoint Theory Reader*.

105 Willey, "World of Materialisms," 1008.

106 Counting and measuring make things. As STS scholar Geoff Bowker (unmarked) writes, "The database itself will ultimately shape the world in its image: it will be performative.

of what your study is gathering knowledge about, a sample is a subset of your population[107] of interest. A statistically good sample–population relationship allows you to generalize from your sample to the population. It allows your count to be of the world, writ small. You can never know if your sample is truly representative of the larger population or phenomenon, but sampling methods exist to make this as likely as possible. One popular sampling method is probability sampling. Probability sampling means that there is a known chance (probability) that each individual instance of the thing you are counting could be selected from the wider population of things. It is the gold standard of reducing count bias and ensuring that your count is counting the thing it says it's counting.[108]

CLEAR does not use probability sampling. Instead, we use judgmental sampling, where researchers actively choose what to include in a count. Our samples are biased (in the statistical sense of the term) by design. They emanate from a particular standpoint (usually called "expertise" in the statistical literature). Judgmental sampling is quite common in pollution science. If someone thinks their oil tank is leaking, they don't grid off their entire yard and randomly select some grid points to sample. This approach might miss the tank entirely but would represent the lawn overall quite well. Instead, they sample around the tank, and often only around the tank. In quantitative research design, that is considered a biased sample. And that's fine in science, so long as everyone knows that's how the count is organized.[109]

If we are only saving what we are counting, and if our counts are skewed in many different ways, then we are creating a new world in which those counts become more and more normalized." Bowker, "Biodiversity Datadiversity," 675. I am here drawing on a rich literature on the politics of counting and measuring, including the way counting makes worlds by determining what is worth counting, how things are categorized for counting (and what does and does not make the cut), and how the metrics of counts impact representations of realities. This includes Nelson, *Who Counts?*; Hacking, "Making and Molding of Child Abuse"; Desrosières and Naish, *Politics of Large Numbers*; Pine and Liboiron, "Politics of Measurement and Action"; Scott, *Seeing like a State*; Verran, *Science and an African Logic*; Verran, "Numbers Performing Nature in Quantitative Valuing"; and Porter, *Trust in Numbers*.

107 Populations, as Michelle Murphy reminds us, are political in part because of how they are created via sampling, counting, and other methods. Murphy, *Economization of Life*.

108 Random sampling is the most common version of probabilistic sampling. It's a methodology that holds that any subject in a population has an identical chance of being selected, so the ones you select are representative of the population.

109 The more science I do, the more I am aware that feminist "strong objectivity" à la Harding is already welcomed and even practiced in dominant scientific culture in various ways.

When we conduct plastic ingestion studies to see whether and to what degree a population of animals ingests plastics, CLEAR uses judgmental sampling by collecting the gastrointestinal tracts ("guts," for short) of animals that are caught for food by humans. Our work does not generalize (or seek to generalize) to animals, but only to the wild food portion[110] of the Newfoundland and Labrador human food web. People here don't eat sperm whales, so CLEAR knows nothing about sperm whales. We know a lot about cod, though. A study design using probabilistic sampling to investigate plastic ingestion rates in fishes would trawl (drag net) fish in a region using a random pattern. That study would be counting plastics in fishes[111] in the area. But we stand on wharfs where fish harvesters land their catch and ask for fish guts. "Hello! I'm a scientist at Memorial University. We're looking for plastics in fish. May I have your carcass when you're done, please?"[112] Because of this protocol, we only sample fish that are likely to be caught by lines and jigs.[113] This means we do not have certain types of fish in our samples, including those that are offshore, smart fish that got away once and no longer fall for lures, fish that have already eaten their dinners, or grandmother fish that tend to be wise and hang out at the bottom of the ocean.[114] We tend to sample the nearby, naive, young, hungry fish that local people eat. We sample freezers, not oceans. Thus, we study food, not fishes.

110 Wild food and country food are terms of art in the province (and elsewhere) that mean food you catch or pick yourself.

111 In science, *fishes* refers to many species of fish. *Fish* is singular for an individual fish, or a population of one species of fish. Trawling gets fishes. Sustenance fishing during cod season gets fish.

112 This is one line of a much longer protocol for gathering fish guts from harvesters. Thanks to CLEAR alum Jessica Melvin (settler), this protocol also includes cutting out tongues, cheeks, and britches (ovaries) from the carcasses we get and offering them back to harvesters as a thank-you for gifting us their guts. These parts are edible and even delicacies but are time-consuming to remove. We then keep the guts for ourselves and throw the rest of the carcass back into the water, where crabs and other life feed on it. We bag and tag the guts, take them back to the lab, and freeze them.

113 Jigging involves dropping a heavy line that has hooks along its length (the jig) into the water and jiggling (jigging) it so the hooks move up and down, snagging fish that swim by. The fish are hooked in the side, not the mouth. It is a method designed for fish-filled waters. I prefer line fishing.

114 Fisheries science now knows that grandmother fish—old females who have spawned often—are the key drivers of fish populations, since they lay more eggs and more of their eggs successfully hatch than younger fish. In fact, overfishing grandmother fish using trawls, which net from the bottom of the ocean, is one of the potential explanations for why the cod fishery collapsed in Newfoundland in 1992. Sustenance fish harvesters' tools

FIGURE 3.4. CLEAR members collecting fish guts at St. Philip's Harbour, Newfoundland. Photo by Bojan Fürst (settler) from the photographic series *How We Do Science* (2018).

Sampling for Sovereignty

We use judgmental sampling because our counts generalize to human food webs. We can thus say things about food sovereignty, even though we can say little about oceans in general. While there is high food insecurity in the province, particularly in parts of Labrador,[115] our anticolonial focus is on food sovereignty (Land relations) in addition to food security (access to wild food). While food security is about access to healthy food, food sovereignty is a broader concept about "the right of peoples to healthy and culturally appropriate food produced through ecologically sound and sustainable methods, and their right to define their own food and agriculture systems."[116] Food sovereignty, more than food se-

are much less likely to hook grandmother fish than industrial bottom trawlers. See Bavington, *Managed Annihilation*.

115 Nunatsiavut Government, "Household Food Security Survey Results Released."

116 Via Campesina, "Food Sovereignty." Via Campesina is an international alliance of organizations of peasant and family farmers, farm workers, Indigenous people, landless

curity, is part of reproductive justice, which includes "how environmental contamination impacts the reproduction of cultural knowledge. . . . At Akwesasne [in southern Ontario], community members report a loss of language and culture around subsistence activities like fishing, which have been largely abandoned because of fears of exposure to contaminants. The generational reproduction of culturally informed interpersonal relationships has been affected as much as physical reproduction. . . . For many indigenous communities, to reproduce culturally informed citizens requires a clean environment."[117]

So, too, in Newfoundland and Labrador. Reproductive justice highlights "the uneven relations and infrastructures that shape what forms of life are supported to persist, thrive, and alter, and what forms of life are destroyed, injured, and constrained,"[118] in this case through traditional food and its contamination.

CLEAR's sampling protocol makes traditional and culturally salient food, rather than strictly fishes, legible as a sink. We ensure the phrase "for human consumption" appears in the titles and keywords of our scientific publications (as well as the name of the place fish were caught).[119] These food webs are not out there in the field, external to either laboratory or daily life. They are our food webs. Some lab members are from Nunatsiavut and NunatuKavut, some are daughters of fish harvesters, and most of us eat local fish (if we eat fish). The lab has a rule about samples: they must be eaten. We do not catch fish for science. We use fish caught for food, and we do science on the leftovers.

---

peasants, and rural women and youth. It is one of the leading actors in global food sovereignty movements.

117 Hoover et al., "Indigenous Peoples of North America," 1648. See chapter 1 for more on pollution and the reproduction of Indigenous culture.

118 Murphy, *Economization of Life*, 141–42.

119 M. Liboiron et al., "Low Plastic Ingestion Rate in Atlantic Cod (*Gadus morhua*) from Newfoundland Destined for Human Consumption Collected through Citizen Science Methods"; M. Liboiron et al., "Low Incidence of Plastic Ingestion among Three Fish Species Significant for Human Consumption on the Island of Newfoundland, Canada"; Saturno et al., "Occurrence of Plastics Ingested by Atlantic Cod (*Gadus morhua*) Destined for Human Consumption (Fogo Island, Newfoundland and Labrador)."

Fish and the other animals we sample are, however, more than food. This section lays out the stakes of articulating cod, for example, as food rather than as a scientific object, but there are also stakes in articulating cod as more than food. Cod are also kin, animals, sentient beings, sovereign bodies, grandmothers, predators, and other things.

FIGURE 3.5. CLEAR's freezer with a box marked "food" to differentiate between guts and edible flesh from sampled fish. There is also a bag of cod skins for Grandmother on the left. Photo by Max Liboiron.

Universalism versus Generalization

Place Doesn't Always Travel

Place-based protocols do not lend themselves to the universal view of dominant science. CLEAR's sampling protocol doesn't always work elsewhere. For instance, on the island of Newfoundland we need fish harvesters to gut their catches on the wharfs so that we can collect the hundreds of gastrointestinal tracts we need. In Labrador, my fishing companions tend to gut the fish halfway home on the rocks on the shoreline, throwing the guts to seals. In Nova Scotia, they tend to take their fish home whole and gut them in the kitchen. Offering someone cod britches (ovaries), cheeks, and tongues as a thank-you isn't received as a gift everywhere. Our protocols—our methods of knowledge production—don't necessarily travel. Elder and CLEAR lab advisor Rick Chavolla (Kumeyaay) says,

> There are certain, very fundamental, elements about a colonial knowledge pursuit in general and it certainly applies to science, maybe in a way even more intensely than almost any other field. One is that there is a universality to it. When you discover something scientifically, it applies to anything, anywhere. You can go anywhere in the world and say, "Yes!

This works! This is what truth is! Truth was here in this place, and truth will be the same someplace else." For us, that's so far from our truth, so far from our knowledge as Indigenous people. We know that for knowledge you *must* understand where you are.[120]

When I first arrived on the island of Newfoundland, I was trained in universal protocols for conducting shoreline microplastic surveys, designed to ensure that scientists can directly compare their findings to one another.[121] The standardized protocol tells you to scoop sand from the top of your quadrat (a little randomly sampled square on the shoreline). But the shorelines here are all rock and ice and snow. They are unscoopable. It is why the island of Newfoundland is called The Rock. Being good sports, we tried using the universal protocol on our rocky coasts but were "unable to procure small microplastics that other literature has demonstrated exists in marine environments"[122] because they either fall through the cracks or are swept back to sea without landing. In short, the universal protocol doesn't work on the island of Newfoundland.[123]

Generalization

If anticolonial sciences eschew universalism in favour of place-based methods, where does that leave the ability for knowledge to work outside of the place of its creation? How do we make a nonuniversal science trustworthy and useful in more than one place? Sometimes it simply does not happen, and that's fine—even good. But usually, even when knowledge is not universal, it is still generalizable. The theory of universalism, where "certain principles, concepts, truths, and values are undeniably valid in all times and places,"[124] tends to overdetermine what generalizable means. Generalizability is about commonality,

120 Richard Chavolla in *Guts*, a documentary directed by Noah Hutton and Taylor Hess. This interview is at 4:23 to 4:51.

121 The universal protocol is here: Lippiatt, Opfer, and Arthur, "Marine Debris Monitoring and Assessment"; EU-TSML (European Union Technical Subgroup on Marine Litter), "Guidance on Monitoring of Marine Litter in European Seas."

122 McWilliams, Liboiron, and Wiersma, "Rocky Shoreline Protocols," 485.

123 The article we published on our failed attempts to use the universal protocol is a strategy to mitigate universalism in science through means that science respects—peer-reviewed publication. Now, I can point to something that shows that universal methods in science do not work, simply and clearly, through proofs and numbers dear to the discipline. For more on how setting precedents in STEM is part of discipline-specific activism, see Gutiérrez and Liboiron, "Strong Animals."

124 Castree, Rogers, and Kitchin, "Universalism."

FIGURE 3.6. I dare you to scoop that. Near Middle Cove, Newfoundland and Labrador. Photo by Max Liboiron.

shared characteristics, and overlap. Things that generalize can still be place-based and have differences, despite similarities.

Tim Choy (unmarked) writes about two moves scholars typically use to deal with the tension between particularism and universalism. One is to subsume the particular into a master narrative, such as the universal We (as discussed in the introduction). The other is to steadfastly refuse the universal, including the project of trying to tie together things that matter at different scales. Choy writes,

> Both responses, whether universalizing or particularizing, seek solid analytic ground; and both find their ground through resort to a "one." This is so whether the one is the unifying one of the "all," or the irreducible particular one refusing subsumption into the general. The conceptual one and the empirical one are a conjoined pair, and both suffer vertigo without firm footing. . . . This might help us to imagine a collective condition that is neither particular nor universal—one governed neither by the "all" nor through the "one nation, one government, one code of laws, one national class-interest, one frontier, and one customs-tariff" that Marx envisioned, nor even the "one planet" of mainstream environmental discourse. Instead, it orients us to the many means, practices, experiences, weather events, and economic relations that co-implicate us at different points as "breathers."[125]

125 Choy, "Air's Substantiations," 11, 12. For more on the various politics of universalism and particularism and how they are mobilized in different ways and simultaneously to different political ends, see Choy, *Ecologies of Comparison*.

He argues that there are ways to think about locality, collectives, and scale that do not "find [their] movement through multiple scales and political forms remarkable in the first place."[126]

Two ways of evaluating whether research is trustworthy and useful in new contexts that attend to multiple scales—that is, the relationships that matter in context—are provocative generalizability and relational validity. Feminist scholar Michelle Fine's (unmarked) concept of provocative generalizability "refers to researchers' attempts to move their findings toward that which is not yet imagined, not yet in practice, not yet in sight. This form of generalizability offers readers an invitation to launch from our findings to what might be, rather than only understanding (or naturalizing) what is."[127] This is an orientation toward an "ought" rather than an "is," a normative orientation. Anticolonial science is an "experimental otherwise"[128] that uses science against scientific values of universalism, separation, domination, and colonization. As such, it is an ideal place for provocative generalizability because it necessitates the creation of new methods and methodologies oriented toward new horizons. It is normative. Getting fish samples from wharfs rather than the ocean and eating our samples allows us not only to talk about food sovereignty, but also to enact it, and thereby allow other researchers to think about how they might also enact food sovereignty, even when our protocols don't exactly work in their place. Publishing papers about how southern "universal" protocols do not work in the north is an invitation for northern-led,[129] place-based methods for shoreline studies that do not require southern landscapes—or scientists.

Another type of generalizability proposed by Tuck and McKenzie is relational validity, which is "based on paradigmatic understandings of the relationality of life."[130] They understand research as a mode of accountability to these relations, an accountability that dominant science has failed: "Ironically, the human induced collapse of ecosystems that has been enabled through non-relational understandings of validity is functioning as a form of the earth 'talking back' in ways that may compel the greater uptake of relational understandings and approaches to legitimacy in research and social life."[131] For example, a sample "disposal" method like CLEAR's that returns guts to the land ac-

126 Choy, "Air's Substantiations," 37.
127 Tuck and McKenzie, *Place in Research*, 229; Fine, "Bearing Witness."
128 Murphy, *Economization of Life*, 105.
129 Moffitt, Chetwynd, and Todd, "Interrupting the Northern Research Industry." Amen.
130 Tuck and McKenzie, *Place in Research*, 157.
131 Tuck and McKenzie, *Place in Research*, 157.

counts for and extends scientists' and fishes' L/land relations and is more valid as an anticolonial method than one that treats all animal tissues as biohazardous waste.

Another way to think about relational validity is how validity is tied to L/land relations and whether or not findings take up the L/land relations that matter in a place. One primary goal of this text has been to centre and nuance land relations in intellectual production so readers can use a form of relational validity in their own work: *how are your actions and research accountable to L/land relations, both colonial and anticolonial? Do your methods, broadly defined, generalize in a way that aligns with the relationalities that are already in play?* I quote Reviewer 2: "The principles [of CLEAR] are replicable even if the [place-based] practices are not. How would scientific practices in the Americas and other colonized regions change if all labs were required to understand what it is to do science in a settler-colonial context—to understand that both the practice of science extends from colonialism and feeds into it?"[132] And how might those changes feed into anticolonial research practices, scientific and otherwise?

Farewell, Good Luck, Generalize

I end on generalization and relational validity both because I think they are hallmarks of anticolonial sciences, but also because you have finished this book and perhaps found parts of it delicious. Perhaps it nourished you. Perhaps you have gobbled up parts and stashed others in the freezer for later. Perhaps it was gross and you spit it out. How might readers relate to this text and its ideas, once we leave the shared page?

Orientations, which come out of obligations, mean that you are facing in a particular direction with a specific horizon of possible action before you. Orientations are the condition of possibility for some futures and not others. Given the diversity of readers and their places, I do not presume to know which direction to point your feet in, but I do know which popular and available orientations reproduce colonial relations to L/land while also sounding like good ideas in academia. If the start of this text was about defamiliarizing and denaturalizing environmental pollution, then the end is about defamiliarizing and denaturalizing reading to make better compromises. What might a reading and citing

132 Reviewer 2, "Reports!," Duke University Press, March 17, 2019. Reviewer 2: Maarsi and much gratitude for these words, but more for keeping me accountable.

practice with an orientation to L/land look like? How do we look out for moves to settler innocence[133] when working across and within difference and how does that change if you are settler, Indigenous, Black, or something else? How do we account for diverse efforts at mitigating colonization from traditions that are not our own, especially if we're Indigenous?

Those weren't rhetorical questions.

This work will look different in different places, which have different relations: "Like colonization, which has shared components and instruments across sites but is uniquely implemented in each setting, decolonization [and anticolonialism] requires unique theories and enactments across sites. Thus, [colonialism] is always historically specific, context specific, and place specific."[134] I hope this book is not used as a Resource, but I do hope the tactic of foregrounding land relations in scientific disciplines to see how they might be "accidentally" colonial and the methodologies of CLEAR's particular anticolonial science generalize relationally and provocatively to your work and obligations. Though I do not have many things figured out, I hope these thoughts have made space for good relations. Maarsi to everyone who made this work possible and as strong as it could be, and to those who will build on it.

That's your cue.

133 You get two conclusions for the price of one. "Conclusion: An ethic of incommensurability, which guides moves that unsettle innocence, stands in contrast to aims of reconciliation, which motivate settler moves to innocence. Reconciliation is about rescuing settler normalcy, about rescuing a settler future. Reconciliation is concerned with questions of what will decolonization look like? What will happen after abolition? What will be the consequences of decolonization for the settler? Incommensurability acknowledges that these questions need not, and perhaps cannot, be answered in order for decolonization to exist as a framework. We want to say, first, that decolonization is not obliged to answer those questions—decolonization is not accountable to settlers, or settler futurity. Decolonization is accountable to Indigenous sovereignty and futurity." Tuck and Yang, "Decolonization Is Not a Metaphor," 35.

134 Tuck and McKenzie, *Place in Research*, 11.

Bibliography

According to Dr. Jane Sumner's (unmarked) Gender Balance Assessment Tool (GBAT), of the 1830 plus authors cited, 38 percent are probably women based on first names.[1] Not good, even after I tried, tried, tried to increase the representation of some authors and remove others. This shows that citational politics[2] needs a method other than "I'd try hard and tweak at the end." CLEAR is working on that.

Adese, Jennifer. "'R' Is for Métis: Contradictions in Scrip and Census in the Construction of a Colonial Métis Identity." *TOPIA: Canadian Journal of Cultural Studies* 25 (2011): 203–12.

Agard-Jones, Vanessa. "Bodies in the System." *Small Axe: A Caribbean Journal of Criticism* 17, no. 3 (2013): 182–92.

Ahmed, Sara. "Making Feminist Points." *Feministkilljoys* (blog), September 11, 2013. https://feministkilljoys.com/2013/09/11/making-feminist-points/.

Ahmed, Sara. *On Being Included: Racism and Diversity in Institutional Life*. Durham, NC: Duke University Press, 2012.

Allen, Edward, Jr. "Neighboring Ontologies: Emerging and Rooted Relationality in Sacred Place." Comprehensive exam, Department of Geography, Memorial University, 2019.

Allen, Tennille, Kellie Carter Jackson, Colin Dayan, Jenny Korn, Danielle Legros Georges, Charles Nfon, Rae Paris, Nicholas Rinehart, Metta Sáma, and Renée Stout. "I Can't Breathe." *Transition: An International Review*, no. 117 (2015): 1–15.

Altman, Rebecca. "American Petro-Topia." *Aeon*, March 11, 2015. https://aeon.co/essays/plastics-run-in-my-family-but-their-inheritance-is-in-us-all.

1 Jane Sumner, "Gender Balance Assessment Tool (GBAT)," accessed August 29, 2020, https://jlsumner.shinyapps.io/syllabustool/.

2 Mott and Cockayne, "Citation Matters"; Ahmed, "Making Feminist Points"; Tuck et al., "Citation Practices"; CBC Radio, "The Politics of Citation"; Erikson and Erlandson, "A Taxonomy of Motives to Cite".

Altman, Rebecca. "Letter to America: Everything Is Going to Have to Be Put Back." *Terrain.Org: A Journal of the Built + Natural Environments*, June 8, 2018. https://www.terrain.org/2018/guest-editorial/letter-to-america-altman/.

Altman, Rebecca. "Time-Bombing the Future." *Aeon*, January 2, 2019. https://aeon.co/essays/how-20th-century-synthetics-altered-the-very-fabric-of-us-all.

Amaral-Zettler, Linda A., Erik R. Zettler, Beth Slikas, Gregory D. Boyd, Donald W. Melvin, Clare E. Morrall, Giora Proskurowski, and Tracy J. Mincer. "The Biogeography of the Plastisphere: Implications for Policy." *Frontiers in Ecology and the Environment* 13, no. 10 (2015): 541–46.

American Chemistry Council. "Chemical Production Expanded in December, 2013 Ended High." *Manufacturing.net*, January 27, 2014. https://www.manufacturing.net/home/article/13150087/chemical-production-expanded-in-december-2013-ended-high.

American Public Health Association, American Water Works Association, and Water Environment Federation. *Standard Methods for the Examination of Water and Wastewater*. 23rd ed. New York: American Public Health Association, 2017. https://www.standardmethods.org/.

American Public Health Association, Laboratory Section of the American Chemistry Society, Association of Official Agricultural Chemists (US), and American Water Works Association. *Standard Methods for the Examination of Water and Sewage*. Vol. 2. New York: American Public Health Association, 1912.

Andersen, Chris. *Métis: Race, Recognition, and the Struggle for Indigenous Peoplehood*. Vancouver: University of British Columbia Press, 2014.

Anderson, Warwick. "Disease, Race, and Empire." *Bulletin of the History of Medicine* 70, no. 1 (1996): 62–67.

Anguksuar, L. R. "A Postcolonial Colonial Perspective on Western [Mis]Conceptions of the Cosmos and the Restoration of Indigenous Taxonomies." In *Two-Spirit People: Native American Gender Identity, Sexuality, and Spirituality*, edited by Sue-Ellen Jacobs, Wesley Thomas, and Sabine Lang, 217–22. Urbana: University of Illinois Press, 1997.

Anker, Peder. *Imperial Ecology: Environmental Order in the British Empire, 1895–1945*. Cambridge, MA: Harvard University Press, 2001.

Anonymous Indigenous Authors. "Indigenization Is Indigenous." *Gazette* (Memorial University), February 12, 2019. https://gazette.mun.ca/campus-and-community/indigenization-is-indigenous/.

Arata, Javier A., Paul R. Sievert, and Maura B. Naughton. *Status Assessment of Laysan and Black-Footed Albatrosses, North Pacific Ocean, 1923–2005*. Reston, VA: US Geological Survey, 2009.

Aris, Aziz. "Estimation of Bisphenol A (BPA) Concentrations in Pregnant Women, Fetuses and Nonpregnant Women in Eastern Townships of Canada." *Reproductive Toxicology* 45 (2014): 8–13.

Armsby, Henry Prentiss. *The Nutrition of Farm Animals*. New York: Macmillan, 1917.

Arnaquq-Baril, Alethea, dir. *Angry Inuk*. Toronto: National Film Board of Canada, 2016.

Aronowsky, Leah. "Gas Guzzling Gaia; or: A Prehistory of Climate Change Denialism." *Critical Inquiry* (forthcoming).

Auman, Heidi J., James P. Ludwig, John P. Giesy, and Theo Colborn. "Plastic Ingestion by Laysan Albatross Chicks on Sand Island, Midway Atoll, in 1994 and 1995." In *Albatross Biology and Conservation*, edited by Graham Robertson and Rosemary Gales, 239–44. London: Surrey Beatty and Sons, 1997.

Avery-Gomm, Stephanie, Michelle Valliant, Carley R. Schacter, Katherine F. Robbins, Max Liboiron, Pierre-Yves Daoust, Lorena M. Rios, and Ian L. Jones. "A Study of Wrecked Dovekies (*Alle alle*) in the Western North Atlantic Highlights the Importance of Using Standardized Methods to Quantify Plastic Ingestion." *Marine Pollution Bulletin* 113, nos. 1–2 (2016): 75–80.

Bäckstrand, Karin. "What Can Nature Withstand? Science, Politics and Discourses in Transboundary Air Pollution Diplomacy." PhD diss., Lund University, 2000.

Bailey-Denton, J. "Sewage Disposal, Ten Years' Experience, with Notes on Sewage Farming." *Activated Sludge Process Development* 49 (1882).

Balayannis, Angeliki, and Emma Garnett. "Chemical Kinship: Interdisciplinary Experiments with Pollution." *Catalyst* 6, no. 1 (2020): 1–10.

Ballestero, Andrea. *A Future History of Water*. Durham, NC: Duke University Press, 2019.

Bang, Megan, Lawrence Curley, Adam Kessel, Ananda Marin, Eli S. Suzukovich III, and George Strack. "Muskrat Theories, Tobacco in the Streets, and Living Chicago as Indigenous Land." *Environmental Education Research* 20, no. 1 (2014): 37–55.

Bang, Megan, and Douglas Medin. "Cultural Processes in Science Education: Supporting the Navigation of Multiple Epistemologies." *Science Education* 94, no. 6 (2010): 1008–26.

Bang, Megan, Douglas Medin, and Gregory Cajete. "Improving Science Education for Native Students: Teaching Place through Community." *Sacnas News* 12, no. 1 (2009): 8–10.

Barad, Karen. *Meeting the Universe Halfway: Quantum Physics and the Entanglement of Matter and Meaning*. Durham, NC: Duke University Press, 2007.

Barman, Jean. "Erasing Indigenous Indigeneity in Vancouver." *BC Studies*, no. 155 (2007): 3.

Bartlett, Cheryl, Murdena Marshall, and Albert Marshall. "Two-Eyed Seeing and Other Lessons Learned within a Co-learning Journey of Bringing Together Indigenous and Mainstream Knowledges and Ways of Knowing." *Journal of Environmental Studies and Sciences* 2, no. 4 (2012): 331–40.

Barton, Gregory Allen. *Empire Forestry and the Origins of Environmentalism*. New York: Cambridge University Press, 2002.

Bashford, Alison. "'Is White Australia Possible?' Race, Colonialism and Tropical Medicine." *Ethnic and Racial Studies* 23, no. 2 (2000): 248–71.

Bavington, Dean. *Managed Annihilation: An Unnatural History of the Newfoundland Cod Collapse*. Vancouver: University of British Columbia Press, 2011.

Beauvoir, Simone de. *The Second Sex*. Translated by Constance Borde and Sheila Malovany-Chevallier. New York: Knopf, 2010.

Bencze, Larry, and Steve Alsop. "Anti-Capitalist/Pro-Communitarian Science and Technology Education." *Journal for Activist Science and Technology Education* 1, no. 1 (2009): 65–84.

Bennett, G. R. "Rubber Bands in a Puffin's Stomach." *British Birds* 53 (1960): 222.

Bergman, Ake, Jerrold J. Heindel, Tim Kasten, Karen A. Kidd, Susan Jobling, Maria Neira, R. Thomas Zoeller, Georg Becher, Poul Bjerregaard, and Riana Bornman. "The Impact of Endocrine Disruption: A Consensus Statement on the State of the Science." *Environmental Health Perspectives* 121, no. 4 (2013): A104–6.

Bhandar, Brenna. *Colonial Lives of Property: Law, Land, and Racial Regimes of Ownership*. Durham, NC: Duke University Press, 2018.

Bijker, Wiebe E. *Of Bicycles, Bakelites, and Bulbs: Toward a Theory of Sociotechnical Change*. Cambridge, MA: MIT Press, 1997.

Bloodgood, Don E. "Water Dilution Factors for Industrial Wastes." *Sewage and Industrial Wastes* (1954): 643–46.

Borrows, John. *Canada's Indigenous Constitution*. Toronto: University of Toronto Press, 2010.

Bowker, Geoffrey C. "Biodiversity Datadiversity." *Social Studies of Science* 30, no. 5 (2000): 643–83.

#breakfreefromplastic. "Green Groups Reveal Top Plastic Polluters Following Massive Beach Cleanup on Freedom Island." *Break Free from Plastic* (blog), December 17, 2017. https://www.breakfreefromplastic.org/2017/12/17/green-groups-reveal-top-plastic-polluters-following-massive-beach-cleanup-on-freedom-island/.

Brighten, Andrew. "Aboriginal Peoples and the Welfare of Animal Persons: Dissolving the Bill C-10B Conflict." *Indigenous Law Journal* 10 (2011): 39.

Brockway, Lucile H. "Science and Colonial Expansion: The Role of the British Royal Botanic Gardens." *American Ethnologist* 6, no. 3 (1979): 449–65.

Brown, Kate. "The Last Sink: The Human Body as the Ultimate Radioactive Storage Site." *RCC Perspectives*, no. 1 (2016): 41–48.

Brynjarsdóttir, Hrönn, and Phoebe Sengers. "Ubicomp from the Edge of the North Atlantic: Lessons from Fishing Villages in Iceland and Newfoundland." In *Ubicomp'09 Workshop*. Orlando, FL: Citeseer, 2009. http://citeseerx.ist.psu.edu/viewdoc/download?doi=10.1.1.505.5772&rep=rep1&type=pdf.

Bullard, Robert D. *Dumping in Dixie: Race, Class, and Environmental Quality*. New York: Routledge, 2018.

Busch, Arthur W. "Use and Abuse of Natural Water Systems." *Journal (Water Pollution Control Federation)* 43, no. 7 (1971): 1480–83.

Bushnik, Tracey, Douglas Haines, Patrick Levallois, Johanne Levesque, Jay Van Oostdam, and Claude Viau. "Lead and Bisphenol A Concentrations in the Canadian Population." *Health Reports* 21, no. 3 (2010): 7–18.

Byrd, Jodi A. *The Transit of Empire: Indigenous Critiques of Colonialism*. Minneapolis: University of Minnesota Press, 2011.

Cairns, John, Jr. "Assimilative Capacity Revisited." *Asian Journal of Experimental Sciences* 22, no. 2 (2008): 177–82.

Cairns, John, Jr. "The Threshold Problem in Ecotoxicology." *Ecotoxicology* 1, no. 1 (1992): 3–16.

Cajete, Gregory. *Native Science: Natural Laws of Interdependence*. Santa Fe, NM: Clear Light Books, 1999.

Calafat, Antonia M., Zsuzsanna Kuklenyik, John A. Reidy, Samuel P. Caudill, John Ekong, and Larry L. Needham. "Urinary Concentrations of Bisphenol A and

4-Nonylphenol in a Human Reference Population." *Environmental Health Perspectives* 113, no. 4 (2004): 391–95.
Calafat, Antonia M., Xiaoyun Ye, Lee-Yang Wong, John A. Reidy, and Larry L. Needham. "Exposure of the US Population to Bisphenol A and 4-Tertiary-Octylphenol: 2003–2004." *Environmental Health Perspectives* 116, no. 1 (2007): 39–44.
Campbell, I. C. "A Critique of Assimilative Capacity." *Journal (Water Pollution Control Federation)* 53, no. 5 (1981): 604–7.
Carroll, Clint. *Roots of Our Renewal: Ethnobotany and Cherokee Environmental Governance*. Minneapolis: University of Minnesota Press, 2015.
Carson, Rachel. *Silent Spring*. Boston: Houghton Mifflin, 2002 [1962].
Castree, Noel, Rob Kitchin, and Alisdair Rogers. "Universalism." In *A Dictionary of Human Geography*. Oxford: Oxford University Press, 2013. https://www.oxfordreference.com/view/10.1093/acref/9780199599868.001.0001/acref-9780199599868-e-1961?rskey=8s3MPx&result=1960.
CBC Radio. "From Scrip to Road Allowances: Canada's Complicated History with the Métis." *Unreserved*, March 28, 2019. https://www.cbc.ca/radio/unreserved/from-scrip-to-road-allowances-canada-s-complicated-history-with-the-métis-1.5100375.
CBC Radio. "The Politics of Citation: Is the Peer Review Process Biased against Indigenous Academics?" *Unreserved*, August 24, 2018. https://www.cbc.ca/radio/unreserved/decolonizing-the-classroom-is-there-space-for-indigenous-knowledge-in-academia-1.4544984/the-politics-of-citation-is-the-peer-review-process-biased-against-indigenous-academics-1.4547468.
Chagani, Fayaz. "Can the Postcolonial Animal Speak?" *Society and Animals* 24, no. 6 (2016): 619–37.
Chakrabarty, Dipesh. "Open Space/Public Place: Garbage, Modernity and India." *South Asia: Journal of South Asian Studies* 14, no. 1 (1991): 15–31.
Chakrabarty, Dipesh. *Provincializing Europe: Postcolonial Thought and Historical Difference*. Princeton, NJ: Princeton University Press, 2009 [2000].
Choy, Timothy. "Air's Substantiations." In *Lively Capital: Biotechnologies, Ethics, and Governance in Global Markets*, edited by Kaushik Sunder Rajan, 121–52. Durham, NC: Duke University Press, 2012.
Choy, Timothy. *Ecologies of Comparison: An Ethnography of Endangerment in Hong Kong*. Durham, NC: Duke University Press, 2011.
Clark, Alfred Joseph. *The Mode of Action of Drugs on Cells*. London: Edward Arnold, 1933.
CLEAR and EDAction. "Pollution Is Colonialism." *Discard Studies* (blog), September 1, 2017. https://discardstudies.com/2017/09/01/pollution-is-colonialism/.
Connell, Raewyn. *Southern Theory: The Global Dynamics of Knowledge in Social Science*. Boston: Polity, 2007.
Conti, Joseph A., and Moira O'Neil. "Studying Power: Qualitative Methods and the Global Elite." *Qualitative Research* 7, no. 1 (2007): 63–82.
Cooley, Oscar W. "Pollution and Property." Foundation for Economic Education (FEE), June 1972. https://fee.org/articles/pollution-and-property/.
Coombs, Amy. "Effects of Exposures on Development of Oocytes: Pat Hunt Q&A." Germline Exposures, March 2014. http://www.germlineexposures.org/pat-hunt-qa.html.

Coulthard, Glen. *Red Skin, White Masks: Rejecting the Colonial Politics of Recognition*. Minneapolis: University of Minnesota Press, 2014.

Cram, Shannon. "Becoming Jane: The Making and Unmaking of Hanford's Nuclear Body." *Environment and Planning D: Society and Space* 33, no. 5 (2015): 796–812.

Crocker, Katherine. "Híyoge Owísisi Tánga Itá (Cricket Egg Stories)." *Carte Blanche*, September 4, 2018. http://carte-blanche.org/hiyoge-owisisi-tanga-ita-cricket-egg-stories/.

Cunsolo, Ashlee, and Karen Landman. *Mourning Nature: Hope at the Heart of Ecological Loss and Grief*. Montreal: McGill-Queen's University Press, 2017.

Curry, Michael. *Digital Places: Living with Geographic Information Technologies*. New York: Routledge, 1998.

Daston, Lorraine. "The History of Science as European Self-Portraiture." *European Review* 14, no. 4 (2006): 523–36.

Daston, Lorraine. "Objectivity versus Truth." *Daimon Revista Internacional de Filosofía*, no. 24 (2001): 11–22.

Davies, Thom. "Slow Violence and Toxic Geographies: 'Out of Sight' to Whom?" *Environment and Planning C: Politics and Space* 0, no. 0 (2019): 1–19.

Davis, Frederick Rowe. *Banned: A History of Pesticides and the Science of Toxicology*. New Haven, CT: Yale University Press, 2014.

Davis, Heather. "Imperceptibility and Accumulation: Political Strategies of Plastic." *Camera Obscura: Feminism, Culture, and Media Studies* 31, no. 2 (2016): 187–93.

Davis, Heather. "Life and Death in the Anthropocene: A Short History of Plastic." In *Art in the Anthropocene: Encounters among Aesthetics, Politics, Environments and Epistemologies*, edited by Heather Davis and Etienne Turpin, 347–58. London: Open Humanities Press, 2015.

Davis, Heather. "Toxic Progeny: The Plastisphere and Other Queer Futures." *PhiloSOPHIA* 5, no. 2 (2015): 231–50.

De Angelis, Massimo. "Marx and Primitive Accumulation: The Continuous Character of Capital's Enclosures." *Commoner* 2, no. 1 (2001): 1–22.

de Coverly, Edd, Pierre McDonagh, Lisa O'Malley, and Maurice Patterson. "Hidden Mountain: The Social Avoidance of Waste." *Journal of Macromarketing* 28, no. 3 (2008): 289–303.

Deloitte and Cheminfo Services. "Economic Study of the Canadian Plastic Industry, Markets, and Waste: Summary Report." Gatineau, QC: Environment and Climate Change Canada, 2019.

De Loughry, Treasa. "Petromodernity, Petro-Finance and Plastic in Karen Tei Yamashita's *Through the Arc of the Rainforest*." *Journal of Postcolonial Writing* 53, no. 3 (2017): 329–41.

De Loughry, Treasa. "Polymeric Chains and Petrolic Imaginaries: World Literature, Plastic, and Negative Value." *Green Letters* 23, no. 2 (2019): 179–93.

Dene Nation and Assembly of First Nations. "We Have Always Been Here: The Significance of Dene Knowledge." Yellowknife, NT: Dene National / Assembly of First Nations Office, 2019.

Denoon, Donald. *Settler Capitalism: The Dynamics of Dependent Development in the Southern Hemisphere*. New York: Oxford University Press, 1983.

Desai, Renu, Colin McFarlane, and Stephen Graham. "The Politics of Open Defecation: Informality, Body, and Infrastructure in Mumbai." *Antipode* 47, no. 1 (2015): 98–120.

Desrosières, Alain, and Camille Naish. *The Politics of Large Numbers: A History of Statistical Reasoning*. Cambridge, MA: Harvard University Press, 2002.

De Wolff, Kim. "Gyre Plastic : Science, Circulation and the Matter of the Great Pacific Garbage Patch." PhD diss., University of California, San Diego, 2014. https://escholarship.org/uc/item/21w9h64q.

De Wolff, Kim. "Plastic Naturecultures: Multispecies Ethnography and the Dangers of Separating Living from Nonliving Bodies." *Body and Society* 23, no. 3 (2017): 23–47.

D'Ignazio, Catherine, and Lauren F. Klein. *Data Feminism*. Cambridge, MA: MIT Press, 2020.

Dillon, Lindsey, and Julie Sze. "Police Powers and Particulate Matters: Environmental Justice and the Spacialities of In/Securities in US Cities." *English Language Notes* 54, no. 2 (2016): 13–23.

Dimaline, Cherie. *The Marrow Thieves*. Toronto: Dancing Cat Books, 2017.

Donald, Dwayne. "Indigenous Métissage: A Decolonizing Research Sensibility." *International Journal of Qualitative Studies in Education* 25, no. 5 (2012): 533–55.

Donald, Dwayne, Florence Glanfield, and Gladys Sterenberg. "Living Ethically within Conflicts of Colonial Authority and Relationality." *Journal of the Canadian Association for Curriculum Studies* 10, no. 1 (2012): 53–76.

Doron, Assa, and Ira Raja. "The Cultural Politics of Shit: Class, Gender and Public Space in India." *Postcolonial Studies* 18, no. 2 (2015): 189–207.

Dosemagen, Shannon, Jeffrey Warren, and Sara Wylie. "Grassroots Mapping: Creating a Participatory Map-Making Process Centered on Discourse." *Journal of Aesthetics and Protest* 8 (2011): 1–11.

Douglas, Mary. *Purity and Danger: An Analysis of Concepts of Pollution and Taboo*. New York: Routledge, 1966.

Douglas, Mary. *Risk and Blame: Essays in Cultural Theory*. London: Routledge, 1992.

Downey, Gary Lee, and Teun Zuiderent-Jerak. "Making and Doing: Engagement and Reflexive Learning in STS." In *The Handbook of Science and Technology Studies*, 3rd ed., edited by Edward J. Hackett, Olga Amsterdamska, Michael E. Lynch, and Judy Wajcman, 223–51. Cambridge, MA: MIT Press, 2007.

Duarte, Marisa Elena. *Network Sovereignty: Building the Internet across Indian Country*. Seattle: University of Washington Press, 2017.

Duarte, Marisa Elena, Morgan Vigil-Hayes, Sandra Littletree, and Miranda Belarde-Lewis. "Of Course, Data Can Never Fully Represent Reality." *Human Biology* 91, no. 3 (2020): 163–78.

Dumit, Joseph. "How I Read." *Dumit* (blog), September 27, 2012. http://dumit.net/how-i-read/.

Dumit, Joseph, Michelle Murphy, Dimitris Papadopoulos, Cori Hayden, and Stefan Helmreich. "Elements Thinking T122.1." Panel presented at the Annual Meeting of the Society for Social Studies of Science, Barcelona, September 3, 2016.

Dunaway, Finis. *Natural Visions: The Power of Images in American Environmental Reform*. Chicago: University of Chicago Press, 2008.

Duncan, Sophie. "Zapatistas Reimagine Science as Tool of Resistance." *Free Radicals* (blog), April 4, 2017. https://freerads.org/2017/04/04/zapatistas-reimagine-science-as-tool-of-resistance/.

Durkalec, A., T. Sheldon, and T. Bell. "Lake Melville: Avativut Kanuittailinnivut (Our Environment, Our Health) Scientific Report." Nain, NL: Nunatsiavut Government, 2016.

Edwards, Paul N. *The Closed World: Computers and the Politics of Discourse in Cold War America*. Cambridge, MA: MIT Press, 1997.

Ellis, Phyllis, dir. *The Country*. Corner Brook, NL; 2018.

Elmore, Bartow J. *Citizen Coke: The Making of Coca-Cola Capitalism*. New York: Norton, 2014.

Erikson, Martin G., and Peter Erlandson. "A Taxonomy of Motives to Cite." *Social Studies of Science* 44, no. 4 (2014): 625–37.

Evans, Meredith. "Becoming Sensor in the Planthroposcene: An Interview with Natasha Myers." Visual and New Media Review, *Fieldsights*, July 9, 2020. https://culanth.org/fieldsights/becoming-sensor-an-interview-with-natasha-myers.

ExxonMobil. "The Sky Is Not Falling." *New York Times*, September 28, 1995.

Fanon, Frantz. "Medicine and Colonialism." In *The Cultural Crisis of Modern Medicine*, edited by John Ehrenreich, 229–51. New York: Monthly Review Press, 1978.

Fennell, Catherine. "The Family Toxic: Triaging Obligation in Post-Welfare Chicago." *South Atlantic Quarterly* 115, no. 1 (2016): 9–32.

Fine, Michelle. "Bearing Witness: Methods for Researching Oppression and Resistance—A Textbook for Critical Research." *Social Justice Research* 19, no. 1 (2006): 83–108.

Fine, Michelle. "Working the Hyphens." In *Handbook of Qualitative Research*, edited by N. K. Denzin and Yvonna S. Lincoln, 70–82. Thousand Oaks, CA: Sage, 1994.

Firth, Barry K. "Status of Water Quality Modeling in the Pulp and Paper Industry." *Journal (Water Pollution Control Federation)* 58, no. 10 (1984): 1131–35.

Fiske, Amelia. "Dirty Hands: The Toxic Politics of Denunciation." *Social Studies of Science* 48, no. 3 (2018): 389–413.

Fortier, Craig. *Unsettling the Commons: Social Movements within, against, and beyond Settler Colonialism*. Winnipeg, MB: ARP Books, 2017.

Fortun, Kim. *Advocacy after Bhopal: Environmentalism, Disaster, New Global Orders*. Chicago: University of Chicago Press, 2009.

Foucault, Michel. *Discipline and Punish: The Birth of the Prison*. 2nd ed. New York: Vintage, 1991.

Friedel, Robert. *Pioneer Plastic: The Making and Selling of Celluloid*. Madison: University of Wisconsin Press, 1983.

Fürst, Bojan. *How We Do Science*. Photographic series, 2018. http://bojanfurstphotography.com/how-we-do-science.

Gabrys, Jennifer, Gay Hawkins, and Mike Michael, eds. *Accumulation: The Material Politics of Plastic*. New York: Routledge, 2013.

Gaztambide-Fernández, Rubén A. "Decolonization and the Pedagogy of Solidarity." *Decolonization: Indigeneity, Education and Society* 1, no. 1 (2012): 41–67.

Gaztambide-Fernández, Rubén A. "Elite Entanglements and the Demand for a Radically Un/Ethical Position: The Case of Wienie Night." *International Journal of Qualitative Studies in Education* 28, no. 9 (2015): 1129–47.
Geniusz, Wendy Makoons. *Our Knowledge Is Not Primitive: Decolonizing Botanical Anishinaabe Teachings*. Syracuse, NY: Syracuse University Press, 2009.
Gerona, Roy R., Tracey J. Woodruff, Carrie A. Dickenson, Janet Pan, Jackie M. Schwartz, Saunak Sen, Matthew W. Friesen, Victor Y. Fujimoto, and Patricia A. Hunt. "Bisphenol-A (BPA), BPA Glucuronide, and BPA Sulfate in Midgestation Umbilical Cord Serum in a Northern and Central California Population." *Environmental Science and Technology* 47, no. 21 (2013): 12477–85.
Geyer, Roland, Jenna R. Jambeck, and Kara Lavender Law. "Production, Use, and Fate of All Plastics Ever Made." *Science Advances* 3, no. 7 (2017): 1–5.
Gibson-Graham, J. K. "The End of Capitalism (as We Knew It): A Feminist Critique of Political Economy." *Capital and Class* 21, no. 2 (1997): 186–88.
Gibson-Graham, J. K. "Rethinking the Economy with Thick Description and Weak Theory." *Current Anthropology* 55, no. S9 (2014): S147–53.
Gien, Lan T. "Land and Sea Connection: The East Coast Fishery Closure, Unemployment and Health." *Canadian Journal of Public Health* 91, no. 2 (2000): 121–24.
Gilio-Whitaker, Dina. *As Long as Grass Grows: The Indigenous Fight for Environmental Justice, from Colonization to Standing Rock*. Boston: Beacon Press, 2019.
Gill, Kaveri. *Of Poverty and Plastic: Scavenging and Scrap Trading Entrepreneurs in India's Urban Informal Economy*. New York: Oxford University Press, 2009.
Gille, Zsuzsa. *From the Cult of Waste to the Trash Heap of History: The Politics of Waste in Socialist and Postsocialist Hungary*. Bloomington: Indiana University Press, 2007.
Glen, Alexander. "Appendix B: Fourth Report of the Commissioners Appointed in 1868 to Inquire into the Best Means of Preventing the Pollution of Rivers." In *The Rivers Pollution Prevention Act, 1876, 39 and 40 Vict. C. 75: With Introduction, Notes, and Index*, 75–79. London: Knight and Company, 1876.
Global Alliance for Incinerator Alternatives Coalition (GAIA). "Open Letter to Ocean Conservancy regarding the Report 'Stemming the Tide.'" October 2015. http://www.no-burn.org/wp-content/uploads/Open_Letter_Stemming_the_Tide_Report_2_Oct_15.pdf.
Global Alliance for Incinerator Alternatives Coalition (GAIA). "Plastics Exposed: How Waste Assessments and Brand Audits Are Helping Philippine Cities Fight Plastic Pollution." Quezon City, Philippines: GAIA, 2009. https://www.no-burn.org/waba2019/.
Goeman, Mishuana. "Land as Life: Unsettling the Logics of Containment." In *Native Studies Keywords*, edited by Stephanie Teves, Andrea Smith, and Michelle Raheja, 71–89. Tucson: University of Arizona Press, 2015.
Goldstein, Jesse. "Terra Economica: Waste and the Production of Enclosed Nature." *Antipode* 45, no. 2 (2013): 357–75.
Government of Canada. "Increasing Knowledge on Plastic Pollution Initiative." Environmental Funding Programs, Government of Canada, March 18, 2020. https://www.canada.ca/en/environment-climate-change/services/environmental-funding/programs/increasing-knowledge-plastic-pollution-initiative.html.

Graham, Mary. "Understanding Human Agency in Terms of Place: A Proposed Aboriginal Research Methodology." *PAN: Philosophy Activism Nature*, no. 6 (2009): 71.

Grande, Sandy. *Red Pedagogy: Native American Social and Political Thought*. Lanham, MD: Rowman and Littlefield, 2015.

Grande, Sandy. "Refusing the University." In *Toward What Justice? Describing Diverse Dreams of Justice in Education*, edited by Eve Tuck and K. Wayne Yang, 57–76. New York: Routledge, 2018.

Grandjean, Philippe. "Paracelsus Revisited: The Dose Concept in a Complex World." *Clinical Pharmacology and Toxicology* 119, no. 2 (2016): 126–32.

Gray-Cosgrove, Carmella, Max Liboiron, and Josh Lepawsky. "The Challenges of Temporality to Depollution and Remediation." *SAPIENS: Surveys and Perspectives Integrating Environment and Society* 8, no. 1 (2015): 1–11.

Grosz, Elizabeth. "Conclusion: A Note on Essentialism and Difference." In *Feminist Knowledge: Critique and Construct*, edited by Sneja Gunew, 332–44. New York: Routledge, 2013.

Grove, Richard. "The Origins of Environmentalism." *Nature* 345, no. 6270 (1990): 11–14.

Gutiérrez, Rochelle, and Max Liboiron. "Strong Animals: Humility in Science." *Science for the People Magazine* 22, no. 2 (2020): 1–4.

Hacking, Ian. "The Making and Molding of Child Abuse." *Critical Inquiry* 17, no. 2 (1991): 253–88.

Hajer, Maarten A. *The Politics of Environmental Discourse: Ecological Modernization and the Policy Process*. Oxford: Clarendon, 1995.

Hale, Charles R. "Activist Research v. Cultural Critique: Indigenous Land Rights and the Contradictions of Politically Engaged Anthropology." *Cultural Anthropology* 21, no. 1 (2006): 96–120.

Hall, Lisa Kahaleole. "Strategies of Erasure: U.S. Colonialism and Native Hawaiian Feminism." *American Quarterly* 60, no. 2 (2008): 273–80.

Hamlin, Christopher. *A Science of Impurity: Water Analysis in Nineteenth Century Britain*. Berkeley: University of California Press, 1990.

Hamlin, Christopher. "'Waters' or 'Water'—Master Narratives in Water History and Their Implications for Contemporary Water Policy." *Water Policy* 2, nos. 4–5 (2000): 313–25.

Hanrahan, Maura. *The Lasting Breach: The Omission of Aboriginal People from the Terms of Union between Newfoundland and Canada and Its Ongoing Impacts*. Research Paper for the Royal Commission on Renewing and Strengthening Our Place in Canada. St. John's: Government of Newfoundland and Labrador, 2003.

Haraway, Donna. "Situated Knowledges: The Science Question in Feminism and the Privilege of Partial Perspective." *Feminist Studies* 14, no. 3 (1988): 575–99.

Harding, Sandra, ed. *The Feminist Standpoint Theory Reader: Intellectual and Political Controversies*. New York: Routledge, 2004.

Harding, Sandra. *Science and Social Inequality: Feminist and Postcolonial Issues*. Urbana: University of Illinois Press, 2006.

Harding, Sandra. *Sciences from Below: Feminisms, Postcolonialities, and Modernities*. Durham, NC: Duke University Press, 2008.

Harris, Cheryl I. "Whiteness as Property." *Harvard Law Review* 106, no. 8 (1993): 1707–91.
Harris, Cole. "How Did Colonialism Dispossess? Comments from an Edge of Empire." *Annals of the Association of American Geographers* 94, no. 1 (2004): 165–82.
Harris, R. Cole. *Making Native Space: Colonialism, Resistance, and Reserves in British Columbia*. Vancouver: University of British Columbia Press, 2011.
Hawkins, Gay. "The Performativity of Food Packaging: Market Devices, Waste Crisis and Recycling." *Sociological Review* 60, no. 2, suppl. (2012): 66–83.
Hawkins, Gay, Emily Potter, and Kane Race. *Plastic Water: The Social and Material Life of Bottled Water*. Cambridge, MA: MIT Press, 2015.
He, Yonghua, Maohua Miao, Lisa J. Herrinton, Chunhua Wu, Wei Yuan, Zhijun Zhou, and De-Kun Li. "Bisphenol A Levels in Blood and Urine in a Chinese Population and the Personal Factors Affecting the Levels." *Environmental Research* 109, no. 5 (2009): 629–33.
Health Canada. "Guidelines for Canadian Drinking Water Quality: Guideline Technical Document—Arsenic." Ottawa, ON: Water Quality and Health Bureau, Healthy Environments and Consumer Safety Branch, Health Canada, May 2006.
Health Canada. "Lead in Drinking Water," March 4, 2016. https://www.canada.ca/en/health-canada/programs/consultation-lead-drinking-water/document.html#a1.
Health Care without Harm. "Health Care without Harm." Accessed May 14, 2019. https://noharm.org/.
Hecht, Gabrielle. "The African Anthropocene." *Aeon*, February 6, 2018. https://aeon.co/essays/if-we-talk-about-hurting-our-planet-who-exactly-is-the-we.
Heglar, Mary Annaïse. "Climate Change Isn't the First Existential Threat." *Medium*, February 18, 2019. https://medium.com/s/story/sorry-yall-but-climate-change-ain-t-the-first-existential-threat-b3c999267aa0.
Heidegger, Martin. *The Question concerning Technology and Other Essays*. Translated by William Lovitt. New York: Harper Torchbooks, 1977.
Helmreich, Stefan. *Alien Ocean: Anthropological Voyages in Microbial Seas*. Berkeley: University of California Press, 2009.
Helmreich, Stefan. "Hokusai's Great Wave Enters the Anthropocene." *Environmental Humanities* 7, no. 1 (2016): 203–17.
Hodges, Sarah. "Medical Garbage and the Making of Neo-liberalism in India." *Economic and Political Weekly* 48, no. 48 (2013): 112–19.
Hoover, Elizabeth. "Cultural and Health Implications of Fish Advisories in a Native American Community." *Ecological Processes* 2, no. 4 (2013): 1–12.
Hoover, Elizabeth. *The River Is in Us: Fighting Toxics in a Mohawk Community*. Minneapolis: University of Minnesota Press, 2017.
Hoover, Elizabeth, Katsi Cook, Ron Plain, Kathy Sanchez, Vi Waghiyi, Pamela Miller, Renee Dufault, Caitlin Sislin, and David O. Carpenter. "Indigenous Peoples of North America: Environmental Exposures and Reproductive Justice." *Environmental Health Perspectives* 120, no. 12 (2012): 1645–49.
Huang, Michelle N. "Ecologies of Entanglement in the Great Pacific Garbage Patch." *Journal of Asian American Studies* 20, no. 1 (2017): 95–117.
Hubbs, Carl L. "Sewage Treatment and Fish Life." *Sewage Works Journal* (1933): 1033–40.

Huff, James. "Carcinogenicity of Bisphenol A Revisited." *Toxicological Sciences* 70, no. 2 (2002): 281–83.

Hunt, Patricia A., Kara E. Koehler, Martha Susiarjo, Craig A. Hodges, Arlene Ilagan, Robert C. Voigt, Sally Thomas, Brian F. Thomas, and Terry J. Hassold. "Bisphenol A Exposure Causes Meiotic Aneuploidy in the Female Mouse." *Current Biology* 13, no. 7 (2003): 546–53.

Hutton, Noah, and Taylor Hess, dirs. *Guts*. Documentary. New York: Couple Three Films, 2019.

Igo, Sarah E. *The Averaged American: Surveys, Citizens, and the Making of a Mass Public*. Cambridge, MA: Harvard University Press, 2007.

Inglis, David. "Dirt and Denigration: The Faecal Imagery and Rhetorics of Abuse." *Postcolonial Studies: Culture, Politics, Economy* 5, no. 2 (2002): 207–21.

Ivie, Eric. "What Do You Mean 'We,' White Man?" *Hunt the Devil* (blog), May 8, 2015. https://huntthedevil.wordpress.com/2015/05/08/what-do-you-mean-we-white-man/.

Jambeck, Jenna R. "Plastic Waste Inputs from Land into the Ocean." *Jambeck Research Group* (blog), 2015. http://jambeck.engr.uga.edu/landplasticinput.

Jambeck, Jenna R., Roland Geyer, Chris Wilcox, Theodore R. Siegler, Miriam Perryman, Anthony Andrady, Ramani Narayan, and Kara Lavender Law. "Plastic Waste Inputs from Land into the Ocean." *Science* 347, no. 6223 (2015): 768–71.

Johnson, Leigh. "The Fearful Symmetry of Arctic Climate Change: Accumulation by Degradation." *Environment and Planning D: Society and Space* 28, no. 5 (2010): 828–47.

Jones, Alison, and Kuni Jenkins. "Rethinking Collaboration: Working the Indigene-Colonizer Hyphen." In *Handbook of Critical Indigenous Methodologies*, edited by Norman K. Denzin, Yvonna S. Lincoln, and Linda Tuhiwai Smith, 471–86. Thousand Oaks, CA: Sage, 2008.

Kao, Shih-yang. "The City Recycled: The Afterlives of Demolished Buildings in Postwar Beijing." PhD thesis, University of California, Berkeley, 2013.

Kawagley, Angayuqaq Oscar. *A Yupiaq Worldview: A Pathway to Ecology and Spirit*. Long Grove, IL: Waveland Press, 2006.

Kawagley, Angayuqaq Oscar, Delena Norris-Tull, and Roger A. Norris-Tull. "The Indigenous Worldview of Yupiaq Culture: Its Scientific Nature and Relevance to the Practice and Teaching of Science." *Journal of Research in Science Teaching: The Official Journal of the National Association for Research in Science Teaching* 35, no. 2 (1998): 133–44.

Keali'ikanaka'oleohaililani, Kekuhi. "Hawaii Environmental Kinship." In *Weaving Indigenous and Sustainability Sciences: Diversifying Our Methods*, edited by Jay T. Johnson, Renee Pualani Louis, and Andrew Kliskey, 77–78. Arlington, VA: National Science Foundation, 2014.

Kenney, Martha. "Fables of Response-Ability: Feminist Science Studies as Didactic Literature." *Catalyst: Feminism, Theory, Technoscience* 5, no. 1 (2019): 1–39.

Kim, Shin Woong, Yooeun Chae, Dokyung Kim, and Youn-Joo An. "Zebrafish Can Recognize Microplastics as Inedible Materials: Quantitative Evidence of Ingestion Behavior." *Science of the Total Environment* 649 (February 1, 2019): 156–62.

Kimelman, Justice E. C. "No Quiet Place: Review Committee on Indian and Métis Adoptions and Placements." Winnipeg, MB: Manitoba Community Services, 1985.

Kimmerer, Robin Wall. *Braiding Sweetgrass: Indigenous Wisdom, Scientific Knowledge, and the Teachings of Plants*. Minneapolis, MN: Milkweed Editions, 2013.
King, Tiffany Lethabo. *The Black Shoals: Offshore Formations of Black and Native Studies*. Durham, NC: Duke University Press, 2019.
Kingsland, Sharon E. *The Evolution of American Ecology, 1890–2000*. Baltimore: Johns Hopkins University Press, 2005.
Klocker, Natascha, Paul Mbenna, and Chris Gibson. "From Troublesome Materials to Fluid Technologies: Making and Playing with Plastic-Bag Footballs." *Cultural Geographies* 25, no. 2 (2018): 301–18.
Knobloch, Frieda. *The Culture of Wilderness: Agriculture as Colonization in the American West*. Chapel Hill: University of North Carolina Press, 1996.
Knudtson, Peter, and David Suzuki. *Wisdom of the Elders: Native and Scientific Ways of Knowing about Nature*. Vancouver: Greystone Books, 2006.
Komeie, Taisaku. "Colonial Environmentalism and Shifting Cultivation in Korea." *Geographical Review of Japan* 79, no. 12 (2006): 664–79.
Kone, Lassana. "Pollution in Africa: A New Toxic Waste Colonialism? An Assessment of Compliance of the Bamako Convention in Cote d'Ivoire." PhD thesis, University of Pretoria, 2009.
Konsmo, Erin Marie, and Karyn Recollet. "Afterword: Meeting the Land(s) Where They Are At." In *Indigenous and Decolonizing Studies in Education: Mapping the Long View*, edited by Linda Tuhiwai Smith, Eve Tuck, and K. Wayne Yang, 238–51. New York: Routledge, 2018.
Kosuth, Mary, Sherri A. Mason, and Elizabeth V. Wattenberg. "Anthropogenic Contamination of Tap Water, Beer, and Sea Salt." *PloS One* 13, no. 4 (2018): e0194970.
Krishnan, Aruna V., Peter Stathis, Suzanne F. Permuth, Laszlo Tokes, and David Feldman. "Bisphenol-A: An Estrogenic Substance Is Released from Polycarbonate Flasks during Autoclaving." *Endocrinology* 132, no. 6 (1993): 2279–86.
Kuhn, Thomas S. *The Structure of Scientific Revolutions*. Chicago: University of Chicago Press, 2012 [1962].
Labrador Research Forum participants. "Caribou and Moose." Presented at the Labrador Research Forum, Happy Valley-Goose Bay, May 2, 2019.
Lagarde, Fabien, Claire Beausoleil, Scott M. Belcher, Luc P. Belzunces, Claude Emond, Michel Guerbet, and Christophe Rousselle. "Non-Monotonic Dose-Response Relationships and Endocrine Disruptors: A Qualitative Method of Assessment." *Environmental Health* 14, no. 1 (2015): 13.
Land, Clare. *Decolonizing Solidarity: Dilemmas and Directions for Supporters of Indigenous Struggles*. London: Zed Books, 2015.
Landecker, Hannah. "Food as Exposure: Nutritional Epigenetics and the New Metabolism." *BioSocieties* 6, no. 2 (2011): 167–94.
Landecker, Hannah. "The Social as Signal in the Body of Chromatin." *Sociological Review* 64, no. 1, suppl. (2016): 79–99.
Landecker, Hannah. "When the Control Becomes the Experiment." *Limn*, no. 3 (2013). https://limn.it/articles/when-the-control-becomes-the-experiment/.

Landecker, Hannah, and Aaron Panofsky. "From Social Structure to Gene Regulation, and Back: A Critical Introduction to Environmental Epigenetics for Sociology." *Annual Review of Sociology* 39, no. 1 (2013): 333–57.

Langston, Nancy. *Toxic Bodies: Hormone Disruptors and the Legacy of DES*. New Haven, CT: Yale University Press, 2010.

Langwick, Stacey Ann. "A Politics of Habitability: Plants, Healing, and Sovereignty in a Toxic World." *Cultural Anthropology* 33, no. 3 (2018): 415–43.

Latour, Bruno. *The Pasteurization of France*. Cambridge, MA: Harvard University Press, 1993.

Latour, Bruno, and Steve Woolgar. *Laboratory Life: The Construction of Scientific Facts*. Princeton, NJ: Princeton University Press, 1979.

Lavers, Jennifer L., and Alexander L. Bond. "Ingested Plastic as a Route for Trace Metals in Laysan Albatross (*Phoebastria immutabilis*) and Bonin Petrel (*Pterodroma hypoleuca*) from Midway Atoll." *Marine Pollution Bulletin* 110, no. 1 (2016): 493–500.

Lavers, Jennifer L., Ian Hutton, and Alexander L. Bond. "Clinical Pathology of Plastic Ingestion in Marine Birds and Relationships with Blood Chemistry." *Environmental Science and Technology* 53, no. 15 (2019): 9224–31.

Lawford, Karen, and Veldon Coburn. "Research, Ethnic Fraud, and the Academy: A Protocol for Working with Indigenous Communities and Peoples." Yellowhead Institute, August 20, 2019. https://yellowheadinstitute.org/2019/08/20/research-ethnic-fraud-and-the-academy-a-protocol-for-working-with-indigenous-communities-and-peoples/.

Lee, Bandy X. "Causes and Cures VII: Structural Violence." *Aggression and Violent Behavior* 28 (2016): 109–14.

Lee, Erica Violet. "I'm Concerned for Your Academic Career If You Talk about This Publicly." *Moontime Warrior* (blog), February 5, 2016. https://moontimewarrior.com/2016/02/05/im-concerned-about-your-academic-career-if-you-talk-about-this-publicly/.

Leopold, Aldo. *Game Management*. Madison: University of Wisconsin Press, 1933.

Lepawsky, Josh. *Reassembling Rubbish: Worlding Electronic Waste*. Cambridge, MA: MIT Press, 2018.

Lerner, Steve. *Sacrifice Zones: The Front Lines of Toxic Chemical Exposure in the United States*. Cambridge, MA: MIT Press, 2010.

Leroux, Darryl. *Distorted Descent: White Claims to Indigenous Identity*. Winnipeg: University of Manitoba Press.

Liboiron, France, Justine Ammendolia, Jacquelyn Saturno, Jessica Melvin, Alex Zahara, Natalie Richárd, and Max Liboiron. "A Zero Percent Plastic Ingestion Rate by Silver Hake (*Merluccius bilinearis*) from the South Coast of Newfoundland, Canada." *Marine Pollution Bulletin* 131 (2018): 267–75.

Liboiron, Max. "Against Awareness, for Scale: Garbage Is Infrastructure, Not Behavior." *Discard Studies* (blog), January 23, 2014. https://discardstudies.com/2014/01/23/against-awareness-for-scale-garbage-is-infrastructure-not-behavior/.

Liboiron, Max. "Care and Solidarity Are Conditions for Interventionist Research." *Engaging Science, Technology, and Society* 2 (2016): 67–72.

Liboiron, Max. "Exchanging." In *Transmissions: Critical Tactics for Making and Communicating Research*, edited by Kat Jungnickel, 89–107. Cambridge, MA: MIT Press, 2020.

Liboiron, Max. "Good Question. Here's Why Treating Plastics as One Kind of Thing in Science and Activism Isn't Useful and Can Even Cause Harm. A Thread." Twitter, @MaxLiboiron, April 30, 2019, 5:42 A.M. https://twitter.com/MaxLiboiron/status/1123160660085477376.

Liboiron, Max. "How Plastic Is a Function of Colonialism." *Teen Vogue*, December 21, 2018. https://www.teenvogue.com/story/how-plastic-is-a-function-of-colonialism.

Liboiron, Max. "Modern Waste as Strategy." *Lo Squaderno: Explorations in Space and Society* 29 (2013): 9–12.

Liboiron, Max. "Not All Marine Fish Eat Plastics." *The Conversation*, July 11, 2018. http://theconversation.com/not-all-marine-fish-eat-plastics-99488.

Liboiron, Max. "The Ocean Conservatory's Call for Mass Incineration in Asia: Disposability for Profit, Fantasies of Containment, and Colonialism." *Discard Studies* (blog), October 3, 2015. https://discardstudies.com/2015/10/03/the-ocean-conservatorys-call-for-mass-incineration-in-asia-disposability-for-profit-fantasies-of-containment-colonialism/.

Liboiron, Max. "Plasticizers: A Twenty-First-Century Miasma." In *Accumulation: The Material Politics of Plastic*, edited by Jennifer Gabrys, Gay Hawkins, and Mike Michael, 22–44. New York: Routledge, 2013.

Liboiron, Max. "The Politics of Measurement: Per Capita Waste and Previous Sewage Contamination." *Discard Studies* (blog), April 22, 2013. https://discardstudies.com/2013/04/22/the-politics-of-measurement-per-capita-waste-and-previous-sewage-contamination/.

Liboiron, Max. "Redefining Pollution and Action: The Matter of Plastics." *Journal of Material Culture* 21, no. 1 (2016): 87–110.

Liboiron, Max. "'Research on Our Own Terms' by Eve Tuck: A Thread by a Listener." CLEAR (blog), May 3, 2019. https://civiclaboratory.nl/2019/05/03/research-on-our-own-terms-by-eve-tuck-a-thread-by-a-listener/.

Liboiron, Max. "Solutions to Waste and the Problem of Scalar Mismatches." *Discard Studies* (blog), February 10, 2014. http://discardstudies.com/2014/02/10/solutions-to-waste-and-the-problem-of-scalar-mismatches/.

Liboiron, Max. "Toxins or Toxicants? Why the Difference Matters." *Discard Studies* (blog), September 11, 2017. https://discardstudies.com/2017/09/11/toxins-or-toxicants-why-the-difference-matters/.

Liboiron, Max. "Waste Colonialism." *Discard Studies* (blog), November 1, 2018. https://discardstudies.com/2018/11/01/waste-colonialism/.

Liboiron, Max, Justine Ammendolia, Katharine Winsor, Alex Zahara, Hillary Bradshaw, Jessica Melvin, Charles Mather, Natalya Dawe, Emily Wells, and France Liboiron. "Equity in Author Order: A Feminist Laboratory's Approach." *Catalyst: Feminism, Theory, Technoscience* 3, no. 2 (2017): 1–17.

Liboiron, Max, France Liboiron, Emily Wells, Natalie Richárd, Alexander Zahara, Charles Mather, Hillary Bradshaw, and Judyannet Murichi. "Low Plastic Ingestion Rate in Atlantic Cod (*Gadus morhua*) from Newfoundland Destined for Human Consumption Collected through Citizen Science Methods." *Marine Pollution Bulletin* 113, nos. 1–2 (2016): 428–37.

Liboiron, Max, Jessica Melvin, Natalie Richárd, Jacquelyn Saturno, Justine Ammendolia, France Liboiron, Louis Charron, and Charles Mather. "Low Incidence of Plastic Ingestion among Three Fish Species Significant for Human Consumption on the Island of Newfoundland, Canada." *Marine Pollution Bulletin* 141 (April 2019): 244–48.

Liboiron, Max, Emily Simmonds, Edward Allen, Emily Wells, Jess Melvin, Alex Zahara, Charles Mather, and All our teachers. "Doing Ethics with Cod." In *STS Making and Doing*, edited by Gary Downey and Teun Zuiderent-Jerak. Cambridge, MA: MIT Press, 2021.

Liboiron, Max, Manuel Tironi, and Nerea Calvillo. "Toxic Politics: Acting in a Permanently Polluted World." *Social Studies of Science* 48, no. 3 (2018): 331–49.

Liboiron, Max, Alex Zahara, and Ignace Schoot. "Community Peer Review: A Method to Bring Consent and Self-Determination into the Sciences." *Preprints.org*, 2018. https://www.preprints.org/manuscript/201806.0104.

Lippiatt, Sherry, Sarah Opfer, and Courtney Arthur. "Marine Debris Monitoring and Assessment: Recommendations for Monitoring Debris Trends in the Marine Environment." Silver Spring, MD: Marine Debris Program, November 2013.

Little Bear, Leroy. "Big Thinking—Leroy Little Bear: Blackfoot Metaphysics 'Waiting in the Wings.'" Presented at the Congress of the Humanities and Social Sciences, Calgary, Alberta, 2016. https://www.youtube.com/watch?v=o_txPA8CiA4.

Locke, John. *Two Treatises of Government*. New York: Cambridge University Press, 1960.

Longino, Helen E. "Can There Be a Feminist Science?" *Hypatia* 2, no. 3 (1987): 51–64.

Lorde, Audre. *The Master's Tools Will Never Dismantle the Master's House*. Berkeley, CA: Crossing Press, 1983.

Lowitt, Kristen. "Examining Fisheries Contributions to Community Food Security: Findings from a Household Seafood Consumption Survey on the West Coast of Newfoundland." *Journal of Hunger and Environmental Nutrition* 8, no. 2 (2013): 221–41.

Lueck, B. F., A. J. Wiley, Ralph H. Scott, and T. F. Wisniewski. "Determination of Stream Purification Capacity." *Sewage and Industrial Wastes* 29, no. 9 (1957): 1054–65.

Lyons, Maryinez. *The Colonial Disease: A Social History of Sleeping Sickness in Northern Zaire, 1900–1940*. New York: Cambridge University Press, 2002.

MacBride, Samantha. "Does Recycling Actually Conserve or Preserve Things?" *Discard Studies* (blog), February 11, 2019. https://discardstudies.com/2019/02/11/12755/.

MacBride, Samantha. *Recycling Reconsidered: The Present Failure and Future Promise of Environmental Action in the United States*. Cambridge, MA: MIT Press, 2011.

MacDonald, David. "Five Reasons the TRC Chose 'Cultural Genocide.'" *Globe and Mail*, July 6, 2015. https://www.theglobeandmail.com/opinion/five-reasons-the-trc-chose-cultural-genocide/article25311423/.

Mai, Lei, Shan-Ni You, Hui He, Lian-Jun Bao, Liang-Ying Liu, and Eddy Y. Zeng. "Riverine Microplastic Pollution in the Pearl River Delta, China: Are Modeled Estimates Accurate?" *Environmental Science and Technology* 53, no. 20 (October 15, 2019): 11810–17.

Manuel, Arthur, and Grand Chief Ronald M. Derrickson. *Unsettling Canada: A National Wake-Up Call*. Toronto: Between the Lines, 2015.

Martin, Aryn, Natasha Myers, and Ana Viseu. "The Politics of Care in Technoscience." *Social Studies of Science* 45, no. 5 (2015): 625–41.

Martin, Emily. *Bipolar Expeditions: Mania and Depression in American Culture*. Princeton, NJ: Princeton University Press, 2007.

Marx, Karl. *Capital*. Vol. 1. Translated by Ben Fowkes. New York: Penguin, 1990 [1976].

Masco, Joseph. "Bad Weather: On Planetary Crisis." *Social Studies of Science* 40, no. 1 (2010): 7–40.

Masco, Joseph. "The Crisis in Crisis." *Current Anthropology* 58, no. S15 (2017): S65–76.

Masco, Joseph. "Interrogating the Threat." Presidential Plenary presented at the Society for Social Studies of Science (4S), Boston, August 30, 2017.

Masco, Joseph. "Side Effect." *Somatosphere*, December 2, 2013. http://somatosphere.net/2013/side-effect.html/.

Masco, Joseph. *The Theater of Operations: National Security Affect from the Cold War to the War on Terror*. Durham, NC: Duke University Press, 2014.

Massachusetts State Board of Health. *Seventh Annual Report of the Board of Education: Together with the Seventh Annual Report of the Secretary of the Board*. Boston, MA: Dutton and Wentworth, 1844.

Massey, Doreen. "A Global Sense of Place." *Marxism Today*, June 1991. http://www.aughty.org/pdf/global_sense_place.pdf.

Mawani, Renisa. "Law as Temporality: Colonial Politics and Indian Settlers." *UC Irvine Law Review* 4 (2014): 65.

McGranahan, Carole. "Theorizing Refusal: An Introduction." *Cultural Anthropology* 31, no. 3 (2016): 319–25.

McGregor, Deborah. "Traditional Ecological Knowledge: An Anishnabe Woman's Perspective." *Atlantis: Critical Studies in Gender, Culture, and Social Justice* 29, no. 2 (2005): 103–9.

McMullin, Juliet. *The Healthy Ancestor: Embodied Inequality and the Revitalization of Native Hawai'ian Health*. Walnut Creek, CA: Left Coast Press, 2010.

McWilliams, Matt, Max Liboiron, and Yolanda Wiersma. "Rocky Shoreline Protocols Miss Microplastics in Marine Debris Surveys (Fogo Island, Newfoundland and Labrador)." *Marine Pollution Bulletin* 129, no. 2 (2018): 480–86.

Meadows, Donella H., Dennis L. Meadows, Jørgen Randers, and William W. Behrens III. "The Limits to Growth: A Report for the Club of Rome." 1972.

Meeker, John D., Sheela Sathyanarayana, and Shanna H. Swan. "Phthalates and Other Additives in Plastics: Human Exposure and Associated Health Outcomes." *Philosophical Transactions of the Royal Society B: Biological Sciences* 364, no. 1526 (2009): 2097–2113.

Meikle, Jeffrey L. *American Plastic: A Cultural History*. New Brunswick, NJ: Rutgers University Press, 1995.

Merchant, Carolyn. *Reinventing Eden: The Fate of Nature in Western Culture*. New York: Routledge, 2004.

Mitchell, Audra, and Aadita Chaudhury. "Worlding beyond 'the' 'End' of 'the World': White Apocalyptic Visions and BIPOC Futurisms." *International Relations* online first (2020). https://journals.sagepub.com/doi/abs/10.1177/0047117820948936.

Moffitt, Morgan, Courtney Chetwynd, and Zoe Todd. "Interrupting the Northern Research Industry: Why Northern Research Should Be in Northern Hands." *Northern Public Affairs* 4, no. 1 (2015). http://www.northernpublicaffairs.ca/index/interrupting-the-northern-research-industry-why-northern-research-should-be-in-northern-hands/.

Monfort, W. F. "A Special Water Standard." *Journal (American Water Works Association)* 2, no. 1 (1915): 65–73.
Moore, Donald S., Jake Kosek, and Anand Pandian. "Introduction: The Cultural Politics of Race and Nature." In *Race, Nature, and the Politics of Difference*, edited by Donald S. Moore, Jake Kosek, and Anand Pandian, 1–70. Durham, NC: Duke University Press, 2003.
Moore, Donald S., Jake Kosek, and Anand Pandian, eds. *Race, Nature, and the Politics of Difference*. Durham, NC: Duke University Press, 2003.
Moreton-Robinson, Aileen. "The Problematics of Identity: Race, Whiteness, and Indigeneity." Panel presentation at the Native American and Indigenous Studies Association, Los Angeles, May 19, 2018.
Moreton-Robinson, Aileen. *The White Possessive: Property, Power, and Indigenous Sovereignty*. Minneapolis: University of Minnesota Press, 2015.
Morgan, Mary S., Margaret Morrison, and Quentin Skinner, eds. *Models as Mediators: Perspectives on Natural and Social Science*. New York: Cambridge University Press, 1999.
Mortimer-Sandilands, Catriona, and Bruce Erickson. *Queer Ecologies: Sex, Nature, Politics, Desire*. Bloomington: Indiana University Press, 2010.
Moten, Fred, and Stefano Harney. *The Undercommons: Fugitive Planning and Black Study*. New York: Minor Compositions, 2013.
Mott, Carrie, and Daniel Cockayne. "Citation Matters: Mobilizing the Politics of Citation toward a Practice of 'Conscientious Engagement.'" *Gender, Place and Culture* 24, no. 7 (2017): 954–73.
MSFD Technical Subgroup on Marine Litter. *Guidance on Monitoring of Marine Litter in European Seas*. Luxembourg: Publications Office of the European Union, 2013.
Mundo, Dawn Marie Elisse. "#breakfreefromplastic Is Supercharging Coastal Cleanups with Brand Audits to Name Corporate Polluters." *Break Free from Plastic* (blog), September 3, 2018.https://www.breakfreefromplastic.org/2018/09/13/brand-audits-to-name-corporate-polluters/.
Murphy, Michelle. "Against Population, towards Alterlife." In *Making Kin Not Population*, edited by Adele E. Clarke and Donna Jeanne Haraway, 101–24. Chicago: Prickly Paradigm Press, 2018.
Murphy, Michelle. "Alterlife and Decolonial Chemical Relations." *Cultural Anthropology* 32, no. 4 (2017): 494–503.
Murphy, Michelle. "Chemical Regimes of Living." *Environmental History* 13, no. 4 (2008): 695–703.
Murphy, Michelle. "Data towards Dismantlement: Activating Data against Environmental Violence and towards Land/Body Relations." Presented at the Native American and Indigenous Studies Association, Hamilton, Aotearoa (New Zealand), June 28, 2019.
Murphy, Michelle. *The Economization of Life*. Durham, NC: Duke University Press, 2017.
Murphy, Michelle. *Sick Building Syndrome and the Problem of Uncertainty: Environmental Politics, Technoscience, and Women Workers*. Durham, NC: Duke University Press, 2006.

Murphy, Michelle. "Unsettling Care: Troubling Transnational Itineraries of Care in Feminist Health Practices." *Social Studies of Science* 45, no. 5 (2015): 717–37.
Murphy, Michelle. "What Can't a Body Do?" Keynote Plenary Lecture presented at the Society for Social Studies of Science (4S), Barcelona, August 31, 2016. https://catalystjournal.org/index.php/catalyst/article/view/28791/html_8.
Myers, John Peterson, Frederick S. vom Saal, Benson T. Akingbemi, Koji Arizono, Scott Belcher, Theo Colborn, Ibrahim Chahoud, et al. "Why Public Health Agencies Cannot Depend on Good Laboratory Practices as a Criterion for Selecting Data: The Case of Bisphenol A." *Environmental Health Perspectives* 117, no. 3 (2009): 309–15.
Myers, Natasha. *Rendering Life Molecular: Models, Modelers, and Excitable Matter*. Durham, NC: Duke University Press, 2015.
Nadasdy, Paul. "The Anti-Politics of TEK: The Institutionalization of Co-Management Discourse and Practice." *Anthropologica* 47, no 2 (2005): 215–32.
Nadasdy, Paul. *Hunters and Bureaucrats: Power, Knowledge, and Aboriginal-State Relations in the Southwest Yukon*. Vancouver: University of British Columbia Press, 2004.
Nadasdy, Paul. "The Politics of TEK: Power and the 'Integration' of Knowledge." *Arctic Anthropology* 18, no. 2 (1999): 1–18.
Nash, Linda. *Inescapable Ecologies: A History of Environment, Disease, and Knowledge*. Berkeley: University of California Press, 2006.
National Toxicology Program. "Carcinogenesis Bioassay of Bisphenol A (CAS No. 80-05-7) in F344 Rats and B6C3F1 Mice (Feed Study)." *National Toxicology Program Technical Report Series* 215. Bethesda, MD: National Toxicology Program, 1982.
Native Youth Sexual Health Network and Women's Earth Alliance. "Violence on the Land, Violence on Our Bodies: Building an Indigenous Response to Environmental Violence." Women's Earth Alliance, 2016. http://landbodydefense.org/uploads/files/VLVBReportToolkit2016.pdf.
Naylor, William. *Trades Waste: Its Treatment and Utilisation. With Special Reference to the Prevention of Rivers Pollution*. London: Charles Griffin and Company, 1902.
Nelson, Diane M. *Who Counts? The Mathematics of Death and Life after Genocide*. Durham, NC: Duke University Press, 2015.
Nelson, Linda R., and Serdar E. Bulun. "Estrogen Production and Action." *Journal of the American Academy of Dermatology* 45, no. 3, suppl. (2001): S116–24.
Neville, Kate J., and Glen Coulthard. "Transformative Water Relations: Indigenous Interventions in Global Political Economies." *Global Environmental Politics* 19, no. 3 (2019): 1–15.
Ngata, Tina, and Max Liboiron. "Māori Plastic Pollution Expertise and Action in Aotearoa." *CLEAR* (blog), July 13, 2020. https://civiclaboratory.nl/2020/07/13/maori-plastic-pollution-expertise-and-action-in-aotearoa/.
Nicoll, Fiona. "Indigenous Sovereignty and the Violence of Perspective: A White Woman's Coming Out Story." *Australian Feminist Studies* 15, no. 33 (2000): 369–86.
Noguchi, Tamao, Kazue Onuki, and Osamu Arakawa. "Tetrodotoxin Poisoning Due to Pufferfish and Gastropods, and Their Intoxication Mechanism." *ISRN Toxicology* 2011 (2011): 1–10.

Novotny, Vladimir, and Peter A. Krenkel. "A Waste Assimilative Capacity Model for a Shallow, Turbulent Stream." *Water Research* 9, no. 2 (1975): 233–41.

Nunatsiavut Government. "Household Food Security Survey Results Released." Nunatsiavut Government, May 23, 2017. Media release. http://www.nunatsiavut.com/wp-content/uploads/2017/05/NEWS-RELEASE-Food-security-survey-results-released.pdf.

Nunatsiavut Government. "Make Muskrat Right." 2016. http://makemuskratright.squarespace.com/.

O'Brien, Mary H. "Being a Scientist Means Taking Sides." *BioScience* 43, no. 10 (1993): 706–8.

Ocean Conservancy. "Stemming the Tide: Land-Based Strategies for a Plastic-Free Ocean." New York: Trash Free Seas Alliance, 2015.

Odum, E. P. *Fundamentals of Ecology*. Philadelphia: Saunders, 1953.

Ofrias, Lindsay. "Invisible Harms, Invisible Profits: A Theory of the Incentive to Contaminate." *Culture, Theory and Critique* 58, no. 4 (2017): 435–56.

Olsén, Lena, Erik Lampa, Detlef A. Birkholz, Lars Lind, and P. Monica Lind. "Circulating Levels of Bisphenol A (BPA) and Phthalates in an Elderly Population in Sweden, Based on the Prospective Investigation of the Vasculature in Uppsala Seniors (PIVUS)." *Ecotoxicology and Environmental Safety* 75 (2012): 242–48.

Olsson, Johanna Alkan. "Setting Limits in Nature and the Metabolism of Knowledge: The Case of the Critical Load Concept." PhD thesis, Linköpings Universitet, 2003.

Ommer, Rosemary E. "After the Moratorium." *Labour / Le Travail* 50 (2002): 395–400.

Osborne, Tracey. "Fixing Carbon, Losing Ground: Payments for Environmental Services and Land (In)Security in Mexico." *Human Geography* 6, no. 1 (2013): 119–33.

Osborne, Tracey, Laurel Bellante, and Nicolena vonHedemann. *Indigenous Peoples and REDD+: A Critical Perspective*. Geneva: Indigenous People's Biocultural Climate Change Assessment Initiative, 2014.

Pacheco-Vega, Raul. "(Re)theorizing the Politics of Bottled Water: Water Insecurity in the Context of Weak Regulatory Regimes." *Water* 11, no. 4 (2019): 658–74.

Packard, Vance. *The Waste Makers*. New York: D. McKay, 1960.

PAME. "Desktop Study on Marine Litter Including Microplastics in the Arctic." Rovaniemi, Finland: Protection of the Arctic Marine Environment, May 2019. https://www.pame.is/images/03_Projects/Arctic_Marine_Pollution/Litter/Desktop_study/Desktop_Study_on_marine_litter.pdf.

paperson, la. "A Ghetto Land Pedagogy: An Antidote for Settler Environmentalism." *Environmental Education Research* 20, no. 1 (2014): 115–30.

paperson, la. *A Third University Is Possible*. Minneapolis: University of Minnesota Press, 2017.

Pasternak, Shiri. *Grounded Authority: The Algonquins of Barriere Lake against the State*. Minneapolis: University of Minnesota Press, 2017.

Pasternak, Shiri. "How Capitalism Will Save Colonialism: The Privatization of Reserve Lands in Canada." *Antipode* 47, no. 1 (2015): 179–96.

Pasternak, Shiri. "Mercenary Colonialism: Third-Party Management." *Ricochet*, October 25, 2017. https://ricochet.media/en/1994.

Pasternak, Shiri. "Transfer Payments Impose Permanent Austerity on Indigenous Communities." *Ricochet*, June 28, 2017. https://ricochet.media/en/1878.

Pathak, Gauri, and Mark Nichter. "The Anthropology of Plastics: An Agenda for Local Studies of a Global Matter of Concern." *Medical Anthropology Quarterly* 33, no. 2 (2019): 307–26.

Pearl, Raymond. *The Biology of Population Growth*. 3rd ed. North Stratford, NH: Ayer Publishing, 1977.

Peck, Jamie. "Explaining (with) Neoliberalism." *Territory, Politics, Governance* 1, no. 2 (2013): 132–57.

Perreault, Tom. "Dispossession by Accumulation? Mining, Water and the Nature of Enclosure on the Bolivian Altiplano." *Antipode* 45, no. 5 (2013): 1050–69.

Phelps, Earle B. "The Chemical Measure of Stream Pollution and Specifications for Sewage Effluents." *American Journal of Public Health* 3, no. 6 (1912): 524–34.

Phelps, Earle B. "Discussion: Domestic and Industrial Wastes in Relation to Public Water Supply: A Symposium." *American Journal of Public Health* 16 (1926): 795–97.

Phelps, Earle B. "Stream Pollution and Its Relation to the Chemical Industries." *Industrial and Engineering Chemistry* 11, no. 10 (1919): 928–29.

Pietsch, Tamson. *Empire of Scholars: Universities, Networks and the British Academic World 1850–1939*. Manchester: University of Manchester Press, 2015.

Pihama, Leonie. "It Seems That Every Day I Get a Request to Meet or Talk with Organisations about How to 'Do' Kaupapa Māori Research Methodology." Thread, Twitter, @kaupapamaori, October 5, 2018, 5:17 A.M. https://twitter.com/kaupapamaori/status/1048140120681902085.

Pine, Kathleen H., and Max Liboiron. "The Politics of Measurement and Action." In *Proceedings of the 33rd Annual ACM Conference on Human Factors in Computing Systems*, 3147–56. Seoul: ACM, 2015.

Pirard, Catherine, Clémence Sagot, Marine Deville, Nathalie Dubois, and Corinne Charlier. "Urinary Levels of Bisphenol A, Triclosan and 4-Nonylphenol in a General Belgian Population." *Environment International* 48 (2012): 78–83.

PlasticsEurope. "Plastics—the Facts 2016: An Analysis of European Plastics Production, Demand and Waste Data." Brussels: PlasticsEurope, 2016.

Plato. *Plato's Phaedrus*. Translated by R. Hackforth. New York: Cambridge University Press, 1993 [1952].

Porter, Theodore M. *Trust in Numbers: The Pursuit of Objectivity in Science and Public Life*. Princeton, NJ: Princeton University Press, 1996.

Povinelli, Elizabeth A. *Economies of Abandonment: Social Belonging and Endurance in Late Liberalism*. Durham, NC: Duke University Press, 2011.

Pratt, Laura A. W. "Decreasing Dirty Dumping—A Reevaluation of Toxic Waste Colonialism and the Global Management of Transboundary Hazardous Waste." *Texas Environmental Law Journal* 41 (2010–11): 147–69.

Price, Jackie. "But You're Inuk, Right?" Presented at the IGOV Indigenous Speaker Series, April 8, 2013. Video. https://www.youtube.com/watch?v=W36cGxXpjWw.

Price, Jennifer. *Flight Maps: Adventures with Nature in Modern America*. New York: Basic Books, 1999.

Provencher, Jennifer F., Alexander L. Bond, Stephanie Avery-Gomm, Stephanie B. Borrelle, Elisa L. Bravo Rebolledo, Sjúrḥur Hammer, Susanne Kühn, Jennifer L. Lavers,

Mark L. Mallory, and Alice Trevail. "Quantifying Ingested Debris in Marine Megafauna: A Review and Recommendations for Standardization." *Analytical Methods* 9, no. 9 (2017): 1454–69.

Puig de la Bellacasa, María. "Matters of Care in Technoscience: Assembling Neglected Things." *Social Studies of Science* 41, no. 1 (2011): 85–106.

Pulido, Laura. "Geographies of Race and Ethnicity II: Environmental Racism, Racial Capitalism and State-Sanctioned Violence." *Progress in Human Geography* 41, no. 4 (2017): 524–33.

Rahman, Proton, Albert Jones, Joseph Curtis, Sylvia Bartlett, Lynette Peddle, Bridget A. Fernandez, and Nelson B. Freimer. "The Newfoundland Population: A Unique Resource for Genetic Investigation of Complex Diseases." *Human Molecular Genetics* 12, no. 2 (2003): R167–72.

Reed, T. V. "Toxic Colonialism, Environmental Justice, and Native Resistance in Silko's *Almanac of the Dead*." *MELUS: Multi-ethnic Literature of the United States* 34, no. 2 (2009): 25–42.

Reisser, Julia, Jeremy Shaw, Gustaaf Hallegraeff, Maira Proietti, David K. A. Barnes, Michele Thums, Chris Wilcox, Britta Denise Hardesty, and Charitha Pattiaratchi. "Millimeter-Sized Marine Plastics: A New Pelagic Habitat for Microorganisms and Invertebrates." *PLoS One* 9, no. 6 (2014): e100289.

Reo, Nicholas J. "The Importance of Belief Systems in Traditional Ecological Knowledge Initiatives." *International Indigenous Policy Journal* 2, no. 4 (2011): 1–4.

Reo, Nicholas James, and Kyle Powys Whyte. "Hunting and Morality as Elements of Traditional Ecological Knowledge." *Human Ecology* 40, no. 1 (2012): 15–27.

Rifkin, Mark. *Beyond Settler Time: Temporal Sovereignty and Indigenous Self-Determination*. Durham, NC: Duke University Press, 2017.

Ritchie, Hannah, and Max Roser. "Plastic Pollution." *Our World in Data*, September 1, 2018. https://ourworldindata.org/plastic-pollution.

Rivers Pollution Commission. "Sixth Report of the Commissioners Appointed in 1868 to Inquire into the Best Means of Preventing the Pollution of Rivers for the Domestic Water Supply of Great Britain." In *The Rivers Pollution Prevention Act, 1876, 39 and 40 Vict. C. 75: With Introduction, Notes, and Index*. London: Knight and Company, 1876.

Roberts, Jody A. "Reflections of an Unrepentant Plastiphobe: Plasticity and the STS Life." *Science as Culture* 19, no. 1 (2010): 101–20.

Robertson, Morgan. "Measurement and Alienation: Making a World of Ecosystem Services." *Transactions of the Institute of British Geographers* 37, no. 3 (2012): 386–401.

Robles-Anderson, Erica, and Max Liboiron. "Coupling Complexity: Ecological Cybernetics as a Resource for Nonrepresentational Moves to Action." In *Sustainable Media: Critical Approaches to Media and Environment*, edited by Nicole Starosielski and Janet Walker, 248–63. New York: Routledge, 2016.

Rochester, Johanna R., and Ashley L. Bolden. "Bisphenol S and F: A Systematic Review and Comparison of the Hormonal Activity of Bisphenol A Substitutes." *Environmental Health Perspectives* 123, no. 7 (2015): 643–50.

Rochman, Chelsea M., Cole Brookson, Jacqueline Bikker, Natasha Djuric, Arielle Earn, Kennedy Bucci, Samantha Athey, et al. "Rethinking Microplastics as a Diverse Contaminant Suite." *Environmental Toxicology and Chemistry* 38, no. 4 (2019): 703–11.

Rochman, Chelsea M., Mark Anthony Browne, Benjamin S. Halpern, Brian T. Hentschel, Eunha Hoh, Hrissi K. Karapanagioti, Lorena M. Rios-Mendoza, Hideshige Takada, Swee Teh, and Richard C. Thompson. "Policy: Classify Plastic Waste as Hazardous." *Nature* 494, no. 7436 (2013): 169–71.

Rochman, Chelsea M., Mark Anthony Browne, A. J. Underwood, Jan A. van Franeker, Richard C. Thompson, and Linda A. Amaral-Zettler. "The Ecological Impacts of Marine Debris: Unraveling the Demonstrated Evidence from What Is Perceived." *Ecology* 97, no. 2 (2016): 302–12.

Rodriguez-Lonebear, Desi. "Building a Data Revolution in Indian Country." In *Indigenous Data Sovereignty: Toward an Agenda*, edited by Tahu Kukutai and John Taylor, 253–73. Acton: Australian National University Press, 2016.

Ross, Loretta. "Treaties, Reconciliation and Me: Public Talk by Loretta Ross, Treaty Commissioner of Manitoba." Public presentation, Memorial University, Newfoundland, September 27, 2018.

Roundtable on Environmental Health Sciences, Research, and Medicine; Board on Population Health and Public Health Practice; and Institute of Medicine. "The Challenge: Chemicals in Today's Society." In *Identifying and Reducing Environmental Health Risks of Chemicals in Our Society: Workshop Summary*. Washington, DC: National Academies Press, 2014. https://www.ncbi.nlm.nih.gov/books/NBK268889/.

Rowlands, J. Craig, Miriam Sander, James S. Bus, and FutureTox Organizing Committee. "FutureTox: Building the Road for 21st Century Toxicology and Risk Assessment Practices." *Toxicological Sciences* 137, no. 2 (2013): 269–77.

Roy, Deboleena. *Molecular Feminisms: Biology, Becomings, and Life in the Lab*. Seattle: University of Washington Press, 2018.

Rudel, Ruthann A., Janet M. Gray, Connie L. Engel, Teresa W. Rawsthorne, Robin E. Dodson, Janet M. Ackerman, Jeanne Rizzo, Janet L. Nudelman, and Julia Green Brody. "Food Packaging and Bisphenol A and Bis(2-Ethyhexyl) Phthalate Exposure: Findings from a Dietary Intervention." *Environmental Health Perspectives* 119, no. 7 (2011): 914.

Rusert, Britt. *Fugitive Science: Empiricism and Freedom in Early African American Culture*. New York: New York University Press, 2017.

Said, Edward W. *Culture and Imperialism*. New York: Vintage, 1993.

Said, Edward W. "Zionism from the Standpoint of Its Victims." *Social Text*, no. 1 (1979): 7–58.

Salmón, Enrique. *Eating the Landscape: American Indian Stories of Food, Identity, and Resilience*. Tucson: University of Arizona Press, 2012.

Saturno, Jacquelyn, Max Liboiron, Justine Ammendolia, Natasha Healey, Elise Earles, Nadia Duman, Ignace Schoot, Tristen Morris, and Brett Favaro. "Occurrence of Plastics Ingested by Atlantic Cod (*Gadus morhua*) Destined for Human Consumption (Fogo Island, Newfoundland and Labrador)." *Marine Pollution Bulletin* 153 (2020): 110993.

Sayre, Nathan. "The Genesis, History, and Limits of Carrying Capacity." *Annals of the Association of American Geographers* 91, no. 1 (2008): 102–34.

Scheinberg, Anne, and Arthur P. J. Mol. "Multiple Modernities: Transitional Bulgaria and the Ecological Modernisation of Solid Waste Management." *Environment and Planning C: Government and Policy* 28, no. 1 (2010): 18–36.

Schiebinger, Londa L. *Nature's Body: Gender in the Making of Modern Science*. New Brunswick, NJ: Rutgers University Press, 2004 [1993].

Schiebinger, Londa, and Claudia Swan, eds. *Colonial Botany: Science, Commerce, and Politics in the Early Modern World*. Philadelphia: University of Pennsylvania Press, 2007.

Schneider, Daniel. *Hybrid Nature: Sewage Treatment and the Contradictions of the Industrial Ecosystem*. Cambridge, MA: MIT Press, 2011.

Schrader, Astrid. "Responding to *Pfiesteria piscicida* (the Fish Killer): Phantomatic Ontologies, Indeterminacy, and Responsibility in Toxic Microbiology." *Social Studies of Science* 40, no. 2 (2010): 275–306.

Schrank, William E., and Noel Roy. "The Newfoundland Fishery and Economy Twenty Years after the Northern Cod Moratorium." *Marine Resource Economics* 28, no. 4 (2013): 397–413.

Schuurman, Nadine, and Geraldine Pratt. "Care of the Subject: Feminism and Critiques of GIS." *Gender, Place and Culture: A Journal of Feminist Geography* 9, no. 3 (2002): 291–99.

Schwartz, Daniel. "Cultural Genocide Label for Residential Schools Has No Legal Implications, Expert Says." *CBC News*, June 13, 2015. https://www.cbc.ca/news/indigenous/cultural-genocide-label-for-residential-schools-has-no-legal-implications-expert-says-1.3110826.

Scott, James C. *Seeing like a State: How Certain Schemes to Improve the Human Condition Have Failed*. New Haven, CT: Yale University Press, 1998.

Sengers, Phoebe. "What I Learned on Change Islands: Reflections on IT and Pace of Life." *Interactions* 18, no. 2 (2011): 40–48.

Seth, Suman. "Putting Knowledge in Its Place: Science, Colonialism, and the Postcolonial." *Postcolonial Studies* 12, no. 4 (2009): 373–88.

Shadaan, Reena, and Michelle Murphy. "Endocrine-Disrupting Chemicals (EDCs) as Industrial and Settler Colonial Structures: Towards a Decolonial Feminist Approach." *Catalyst: Feminism, Theory, Technoscience* 6, no. 1 (2020): 1–36.

Shapiro, Nicholas. "Attuning to the Chemosphere: Domestic Formaldehyde, Bodily Reasoning, and the Chemical Sublime." *Cultural Anthropology* 30, no. 3 (2015): 368–93.

Shapiro, Nicholas, Nasser Zakariya, and Jody Roberts. "A Wary Alliance: From Enumerating the Environment to Inviting Apprehension." *Engaging Science, Technology, and Society* 3 (2017): 575–602.

Shotwell, Alexis. *Against Purity: Living Ethically in Compromised Times*. Minneapolis: University of Minnesota Press, 2016.

Sileo, Louis, Paul R. Sievert, and Michael D. Samuel. "Causes of Mortality of Albatross Chicks at Midway Atoll." *Journal of Wildlife Diseases* 26, no. 3 (1990): 329–38.

Simmons, Kristen. "Settler Atmospherics—Cultural Anthropology." *Cultural Anthropology* (blog), November 20, 2017. https://culanth.org/fieldsights/1221-settler-atmospherics.

Simpson, Audra. "Consent's Revenge." *Cultural Anthropology* 31, no. 3 (2016): 326–33.

Simpson, Audra. "On Ethnographic Refusal: Indigeneity, 'Voice' and Colonial Citizenship." *Junctures: The Journal for Thematic Dialogue*, no. 9 (2007): 67–80.

Simpson, Audra. "The Ruse of Consent and the Anatomy of 'Refusal': Cases from Indigenous North America and Australia." *Postcolonial Studies* 20, no. 1 (2017): 18–33.

Simpson, Leanne Betasamosake. *Dancing on Our Turtle's Back: Stories of Nishnaabeg Re-Creation, Resurgence and a New Emergence*. Winnipeg, MB: Arbeiter Ring Publishing, 2011.

Smith, Linda Tuhiwai. *Decolonizing Methodologies: Research and Indigenous Peoples*. London: Zed, 1999.

Somsen, Geert J. "A History of Universalism: Conceptions of the Internationality of Science from the Enlightenment to the Cold War." *Minerva* 46, no. 3 (2008): 361–79.

Sontag, Susan. *Against Interpretation and Other Essays*. New York: Macmillan, 2001.

Spackman, Christy, and Gary A. Burlingame. "Sensory Politics: The Tug-of-War between Potability and Palatability in Municipal Water Production." *Social Studies of Science* 48, no. 3 (2018): 350–71.

Stamatopoulou-Robbins, Sophia. "An Uncertain Climate in Risky Times: How Occupation Became like the Rain in Post-Oslo Palestine." *International Journal of Middle East Studies* 50, no. 3 (2018): 383–404.

Stanes, Elyse, and Chris Gibson. "Materials That Linger: An Embodied Geography of Polyester Clothes." *Geoforum* 85 (October 1, 2017): 27–36.

Staniforth, Jesse. "'Cultural Genocide'? No, Canada Committed Regular Genocide." *Star* (Toronto), June 10, 2015. https://www.thestar.com/opinion/commentary/2015/06/10/cultural-genocide-no-canada-committed-regular-genocide.html.

Star, Susan Leigh. "Power, Technology and the Phenomenology of Conventions: On Being Allergic to Onions." *Sociological Review* 38, no. S1 (1990): 26–56.

Stengers, Isabelle. *Another Science Is Possible: A Manifesto for Slow Science*. Translated by Stephen Muecke. Cambridge: Polity, 2018.

Stouffer, Lloyd. "Plastics Packaging: Today and Tomorrow." In *National Plastics Conference*. New York: Society for the Plastics Industry, 1963.

Strakosch, Elizabeth, and Alissa Macoun. "The Vanishing Endpoint of Settler Colonialism." *Arena Journal*, nos. 37–38 (2012): 40–62.

Strasser, Susan. *Waste and Want: A Social History of Trash*. New York: Metropolitan, 1999.

Streeter, H. W., and E. B. Phelps. *A Study of the Pollution and Natural Purification of the Ohio River*. Public Health Service Bulletin 146. Washington, DC: US Department of Health, Education, and Welfare, 1925.

Styres, Sandra, and Dawn Zinga. "The Community-First Land-Centred Theoretical Framework: Bringing a 'Good Mind' to Indigenous Education Research?" *Canadian Journal of Education* 36, no. 2 (2013): 284–313.

Subramaniam, Banu. *Ghost Stories for Darwin: The Science of Variation and the Politics of Diversity*. Urbana: University of Illinois Press, 2014.

Subramaniam, Banu, and Angela Willey. "Introduction: Feminism's Sciences." *Catalyst: Feminism, Theory, Technoscience* 3, no. 1 (2017): 1–23.

Sumner, Jane. "Gender Balance Assessment Tool (GBAT)." Accessed October 21, 2019. https://jlsumner.shinyapps.io/syllabustool/.

Susiarjo, Martha, Terry J. Hassold, Edward Freeman, and Patricia A. Hunt. "Bisphenol A Exposure In Utero Disrupts Early Oogenesis in the Mouse." *PLoS Genetics* 3, no. 1 (2007): E5.

TallBear, Kim. "Indigenous Bioscientists Constitute Knowledge across Cultures of Expertise and Tradition: An Indigenous Standpoint Research Project." In *Re: Mindings: Co-Constituting Indigenous / Academic / Artistic Knowledges*, edited by Johan Gardebo, May-Britt Ohman, and Hiroshi Maruyama, 173–91. Uppsala University, Hugo Valenton Centre, 2014.

TallBear, Kim. "Standing with and Speaking as Faith: A Feminist-Indigenous Approach to Inquiry." *Journal of Research Practice* 10, no. 2 (2014): 1–7.

TallBear, Kim, Candis Callison, and Rick Harp. "Political Pundits' Push-Back on 'Protectors.'" *MediaIndigena*, episode 198, February 24, 2020. https://mediaindigena.libsyn.com/ep-198-political-pundits-push-back-on-protectors.

Tarr, Joel A. "Industrial Wastes and Public Health: Some Historical Notes, Part I, 1876–1932." *American Journal of Public Health* 75, no. 9 (1985): 1059–67.

Tarr, Joel A. "Searching for a 'Sink' for an Industrial Waste: Iron-Making Fuels and the Environment." *Environmental History Review* 18, no. 1 (1994): 9–34.

Thorpe, Charles, and Ian Welsh. "Beyond Primitivism: Toward a Twenty-First Century Anarchist Theory and Praxis for Science." *Anarchist Studies* 16 (2008): 48–75.

Todd, Zoe. "Fish, Kin and Hope: Tending to Water Violations in Amiskwaciwâskahikan and Treaty Six Territory." *Afterall: A Journal of Art, Context and Enquiry* 43, no. 1 (2017): 102–7.

Todd, Zoe. "From Fish Lives to Fish Law: Learning to See Indigenous Legal Orders in Canada." *Somatosphere*, February 1, 2016. http://somatosphere.net/2016/02/from-fish-lives-to-fish-law-learning-to-see-indigenous-legal-orders-in-canada.html.

Todd, Zoe. "An Indigenous Feminist's Take on the Ontological Turn: 'Ontology' Is Just Another Word for Colonialism." *Journal of Historical Sociology* 29, no. 1 (2016): 4–22.

Todd, Zoe. "Refracting Colonialism in Canada: Fish Tales, Text, and Insistent Public Grief." In *Coloniality, Ontology, and the Question of the Posthuman*, edited by Mark Jackson, 147–62. New York: Routledge, 2017.

Todd, Zoe. "Refracting the State through Human-Fish Relations: Fishing, Indigenous Legal Orders and Colonialism in North/Western Canada." *Decolonization: Indigeneity, Education and Society* 7, no. 1 (2018): 60–75.

Tough, Frank, and Erin McGregor. "'The Rights to the Land May Be Transferred': Archival Records as Colonial Text—A Narrative of Métis Scrip." *Canadian Review of Comparative Literature / Revue Canadienne de Littérature Comparée* 34, no. 1 (2011): 33–63.

Trask, Haunani-Kay. *From a Native Daughter: Colonialism and Sovereignty in Hawai'i.* Honolulu: University of Hawai'i Press, 1993.

Trask, Haunani-Kay. "The Color of Violence." In *The Color of Violence*, edited by INCITE! Women of Color against Violence, 82–87. Boston: South End Press, 2006.

Truth and Reconciliation Commission of Canada. *Honouring the Truth, Reconciling for the Future: Summary of the Final Report of the Truth and Reconciliation Commission of Canada*. Ottawa, ON: Lorimer, 2015.

Tuck, Eve. "Research on Our Own Terms." Presented at the Labrador Research Forum, North West River, Labrador, May 1, 2019.

Tuck, Eve. "Suspending Damage: A Letter to Communities." *Harvard Educational Review* 79, no. 3 (2009): 409–28.

Tuck, Eve. "To Watch the White Settlers Sift through Our Work as They Ask, 'Isn't There More for Me Here?'" Thread, Twitter, @tuckeve, October 8, 2017, 12:40 P.M. https://twitter.com/tuckeve/status/917068205667110912.

Tuck, Eve, and Marcia McKenzie. *Place in Research: Theory, Methodology, and Methods.* New York: Routledge, 2015.

Tuck, Eve, and K. Wayne Yang. "Decolonization Is Not a Metaphor." *Decolonization: Indigeneity, Education and Society* 1, no. 1 (2012): 1–40.

Tuck, Eve, and K. Wayne Yang. "R-Words: Refusing Research." In *Humanizing Research: Decolonizing Qualitative Inquiry with Youth and Communities*, edited by Django Paris and Maisha T. Winn, 223–48. Thousand Oaks, CA: Sage, 2014.

Tuck, Eve, and K. Wayne Yang. "Unbecoming Claims: Pedagogies of Refusal in Qualitative Research." *Qualitative Inquiry* 20, no. 6 (2014): 811–18.

Tuck, Eve, K. Wayne Yang, and Rubén Gaztambide-Fernández. "Citation Practices." *Critical Ethnic Studies* (blog), April 2015. http://www.criticalethnicstudiesjournal.org/citation-practices.

Underkuffler, Laura S. *The Idea of Property: Its Meaning and Power*. New York: Oxford University Press, 2003.

United Nations Environment Programme. "Global Chemicals Outlook II: From Legacies to Innovative Solutions." Nairobi: United Nations Environment Programme, 2019.

United States Public Health Service. "Public Health Service Drinking Water Standards." *Journal (American Water Works Association)* (1943): 93–104.

Ureta, Sebastian. "Caring for Waste: Handling Tailings in a Chilean Copper Mine." *Environment and Planning A: Economy and Space* 48, no. 8 (2016): 1532–48.

US Environmental Protection Agency. "Bisphenol A. (CASRN)." Washington, DC: EPA Integrated Risk Information System, 1988. http://www.epa.gov/iris/subst/0356.htm.

US Environmental Protection Agency. "TSCA Chemical Substance Inventory." Collections and Lists. TSCA Chemical Substance Inventory: Basic Information, August 15, 2014. https://www.epa.gov/tsca-inventory.

van Anders, Sari. "Van Anders Lab." 2019. https://www.queensu.ca/psychology/van-anders-lab/.

Vandenberg, Laura N. "Low-Dose Effects of Hormones and Endocrine Disruptors." In *Endocrine Disrupters*, Vol. 94 of *Vitamins and Hormones*, edited by Gerald Litwack, 129–65. New York: Elsevier Academic Press, 2014.

Vandenberg, Laura N., Ibrahim Chahoud, Vasantha Padmanabhan, Francisco J. R. Paumgartten, and Gilbert Schoenfelder. "Biomonitoring Studies Should Be Used by Regulatory Agencies to Assess Human Exposure Levels and Safety of Bisphenol A." *Environmental Health Perspectives* 118, no. 8 (2010): 1051–54.

Vandenberg, Laura N., Theo Colborn, Tyrone B. Hayes, Jerrold J. Heindel, David R. Jacobs Jr., Duk-Hee Lee, John Peterson Myers, et al. "Regulatory Decisions on Endocrine Disrupting Chemicals Should Be Based on the Principles of Endocrinology." *Reproductive Toxicology* 38 (2013): 1–15.

Vandenberg, Laura N., Theo Colborn, Tyrone B. Hayes, Jerrold J. Heindel, David R. Jacobs Jr., Duk-Hee Lee, Toshi Shioda, et al. "Hormones and Endocrine-Disrupting

Chemicals: Low-Dose Effects and Nonmonotonic Dose Responses." *Endocrine Reviews* 33, no. 3 (2012): 378–455.
Vandenberg, Laura N., Derek Luthi, and D'Andre Quinerly. "Plastic Bodies in a Plastic World: Multi-Disciplinary Approaches to Study Endocrine Disrupting Chemicals." *Journal of Cleaner Production* 140, no. 1 (2017): 373–85.
Vandenberg, Laura N., Maricel V. Maffini, Carlos Sonnenschein, Beverly S. Rubin, and Ana M. Soto. "Bisphenol-A and the Great Divide: A Review of Controversies in the Field of Endocrine Disruption." *Endocrine Reviews* 30, no. 1 (2009): 75–95.
van Franeker, J. A., M. Heubeck, K. Fairclough, D. M. Turner, M. Grantham, E. W. M. Stienen, N. Guse, J. Pedersen, K. O. Olsen, and P. J. Andersson. "'Save the North Sea' Fulmar Study 2002–2004: A Regional Pilot Project for the Fulmar-Litter-EcoQO in the OSPAR Area." Wageningen: Alterra, 2005.
Velz, C. J. "Utilization of Natural Purification Capacity in Sewage and Industrial Waste Disposal." *Sewage and Industrial Wastes* 22, no. 12 (1950): 1601–13.
Veracini, Lorenzo. *The Settler Colonial Present*. New York: Palgrave Macmillan, 2015.
Verran, Helen. "Numbers Performing Nature in Quantitative Valuing." *NatureCulture* 2 (2013): 23–37.
Verran, Helen. *Science and an African Logic*. Chicago: University of Chicago Press, 2001.
Via Campesina. "Food Sovereignty." *Via Campesina*, January 15, 2003. https://viacampesina.org/en/food-sovereignty/.
Vizenor, Gerald Robert. *Manifest Manners: Narratives on Postindian Survivance*. Lincoln: University of Nebraska Press, 1999.
Vogel, Sarah Ann. *Is It Safe? BPA and the Struggle to Define the Safety of Chemicals*. Berkeley: University of California Press, 2013.
Vogel, Sarah A. "The Politics of Plastics: The Making and Unmaking of Bisphenol A 'Safety.'" *American Journal of Public Health* 99, no. S3 (2009): S559–66.
vom Saal, Frederick S., Benson T. Akingbemi, Scott M. Belcher, Linda S. Birnbaum, D. Andrew Crain, Marcus Eriksen, Francesca Farabollini, et al. "Chapel Hill Bisphenol A Expert Panel Consensus Statement: Integration of Mechanisms, Effects in Animals and Potential to Impact Human Health at Current Levels of Exposure." *Reproductive Toxicology* 24, no. 2 (2007): 131–38.
vom Saal, Frederick S., and Claude Hughes. "An Extensive New Literature Concerning Low-Dose Effects of Bisphenol A Shows the Need for a New Risk Assessment." *Environmental Health Perspectives* 113, no. 8 (2005): 926–33.
Vowel, Chelsea. *Indigenous Writes: A Guide to First Nations, Métis, and Inuit Issues in Canada*. Winnipeg, MB: Portage and Main Press, 2016.
Voyles, Traci Brynne. *Wastelanding: Legacies of Uranium Mining in Navajo Country*. Minneapolis: University of Minnesota Press, 2015.
Wadiwel, Dinesh Joseph. "Do Fish Resist?" *Cultural Studies Review* 22, no. 1 (2016): 196.
Wagner-Lawlor, Jennifer. "Poor Theory and the Art of Plastic Pollution in Nigeria: Relational Aesthetics, Human Ecology, and 'Good Housekeeping.'" *Social Dynamics* 44, no. 2 (2018): 198–220.
Walia, Harsha. "Decolonizing Together: Moving beyond a Politics of Solidarity toward a Practice of Decolonization." *Briarpatch Magazine* 1 (2012): n.p.

Walker, J. Samuel. *Permissible Dose: A History of Radiation Protection in the Twentieth Century*. Berkeley: University of California Press, 2000.

Walter, Maggie, and Chris Andersen. *Indigenous Statistics: A Quantitative Research Methodology*. Walnut Creek, CA: Left Coast Press, 2013.

Watts, Vanessa. "Indigenous Place-Thought and Agency amongst Humans and Non-Humans (First Woman and Sky Woman Go on a European World Tour!)." *Decolonization: Indigeneity, Education and Society* 2, no. 1 (2013): 20–34.

Westermann, Andrea. "When Consumer Citizens Spoke Up: West Germany's Early Dealings with Plastic Waste." *Contemporary European History* 22, no. 3 (2013): 477–98.

Westman, Walter E. "Some Basic Issues in Water Pollution Control Legislation: Contrasts between Technological and Ecological Perspectives on the Regulation of Effluents Underlie Current Debates in Water Pollution Legislation." *American Scientist* 60, no. 6 (1972): 767–73.

Whitt, Laurelyn. *Science, Colonialism, and Indigenous Peoples: The Cultural Politics of Law and Knowledge*. New York: Cambridge University Press, 2009.

Whyte, Kyle. "The Dakota Access Pipeline, Environmental Injustice, and U.S. Colonialism." *Red Ink: An International Journal of Indigenous Literature, Arts, and Humanities* 19, no. 1 (2017): 154–69.

Whyte, Kyle. "Indigenous Experience, Environmental Justice and Settler Colonialism." SSRN Scholarly Paper. Rochester, NY: Social Science Research Network, April 25, 2016.

Whyte, Kyle Powys. "Is It Colonial Déjà Vu? Indigenous Peoples and Climate Injustice." In *Humanities for the Environment: Integrating Knowledge, Forging New Constellations of Practice*, edited by Joni Adamson and Michael Davis, 102–19. New York: Routledge, 2016.

Whyte, Kyle Powys, Joseph P. Brewer, and Jay T. Johnson. "Weaving Indigenous Science, Protocols and Sustainability Science." *Sustainability Science* 11, no. 1 (2016): 25–32.

Willey, Angela. *Undoing Monogamy: The Politics of Science and the Possibilities of Biology*. Durham, NC: Duke University Press, 2016.

Willey, Angela. "A World of Materialisms: Postcolonial Feminist Science Studies and the New Natural." *Science, Technology, and Human Values* 41, no. 6 (2016): 991–1014.

Wilson, Alex. "N'tacinowin Inna Nah': Our Coming In Stories." *Canadian Woman Studies* 26, no. 3 (2008): 193–99.

Wilson, Shawn. *Research Is Ceremony: Indigenous Research Methods*. Black Point, NS: Fernwood Publishing, 2008.

Wolfe, Patrick. *Settler Colonialism and the Transformation of Anthropology: The Politics and Poetics of an Ethnographic Event*. London: Cassell, 1999.

Wong, Alice. "The Rise and Fall of the Plastic Straw: Sucking in Crip Defiance." *Catalyst: Feminism, Theory, Technoscience* 5, no. 1 (2019): 1–12.

Work, T. M., M. R. Smith, and R. Duncan. "Necrotizing Enteritis as a Cause of Mortality in Laysan Albatross, Diomedea immutabilis, Chicks on Midway Atoll, Hawaii." *Avian Diseases* 41, no. 2 (1998): 1–5.

World Health Organization and United Nations Environment Programme. "State of the Science of Endocrine Disrupting Chemicals." WHO/UNEP, 2012. http://www.who.int/ceh/publications/endocrine/en/.

Wylie, Sara Ann. *Fractivism: Corporate Bodies and Chemical Bonds*. Durham, NC: Duke University Press, 2018.

Wylie, Sara, Nick Shapiro, and Max Liboiron. "Making and Doing Politics through Grassroots Scientific Research on the Energy and Petrochemical Industries." *Engaging Science, Technology, and Society* 3 (2017): 393–425.

Zahara, Alex. "Refusal as Research Method in Discard Studies." *Discard Studies* (blog), March 21, 2016. https://discardstudies.com/2016/03/21/refusal-as-research-method-in-discard-studies/.

Zara. "I Don't Know Who Needs to Hear This Right Now." Twitter, @zaranosaur, 8:08 a.m., July 21, 2019. https://twitter.com/zaranosaur/status/1152913323995869184.

Zettler, Erik R., Tracy J. Mincer, and Linda A. Amaral-Zettler. "Life in the 'Plastisphere': Microbial Communities on Plastic Marine Debris." *Environmental Science and Technology* 47, no. 13 (2013): 7137–46.

Zilberstein, Anya. *A Temperate Empire: Making Climate Change in Early America*. New York: Oxford University Press, 2016.

Index